T0258856

GENETIC ALGORITHMS

for

PATTERN RECOGNITION

Edited by

Sankar K. Pal • Paul P. Wang

CRC Press
Taylor & Francis Group
Boca Raton London New York

CRC Press is an imprint of the
Taylor & Francis Group, an **informa** business

First published 1996 by CRC Press
Taylor & Francis Group
6000 Broken Sound Parkway NW, Suite 300
Boca Raton, FL 33487-2742

Reissued 2018 by CRC Press

© 1996 by Taylor & Francis
CRC Press is an imprint of Taylor & Francis Group, an Informa business

No claim to original U.S. Government works

This book contains information obtained from authentic and highly regarded sources. Reasonable efforts have been made to publish reliable data and information, but the author and publisher cannot assume responsibility for the validity of all materials or the consequences of their use. The authors and publishers have attempted to trace the copyright holders of all material reproduced in this publication and apologize to copyright holders if permission to publish in this form has not been obtained. If any copyright material has not been acknowledged please write and let us know so we may rectify in any future reprint.

Except as permitted under U.S. Copyright Law, no part of this book may be reprinted, reproduced, transmitted, or utilized in any form by any electronic, mechanical, or other means, now known or hereafter invented, including photocopying, microfilming, and recording, or in any information storage or retrieval system, without written permission from the publishers.

For permission to photocopy or use material electronically from this work, please access www.copyright.com (http://www.copyright.com/) or contact the Copyright Clearance Center, Inc. (CCC), 222 Rosewood Drive, Danvers, MA 01923, 978-750-8400. CCC is a not-for-profit organiza-tion that provides licenses and registration for a variety of users. For organizations that have been granted a photocopy license by the CCC, a separate system of payment has been arranged.

Trademark Notice: Product or corporate names may be trademarks or registered trademarks, and are used only for identification and explanation without intent to infringe.

A Library of Congress record exists under LC control number: 95046195

Publisher's Note
The publisher has gone to great lengths to ensure the quality of this reprint but points out that some imperfections in the original copies may be apparent.

Disclaimer
The publisher has made every effort to trace copyright holders and welcomes correspondence from those they have been unable to contact.

ISBN 13: 978-1-138-10557-7 (hbk)
ISBN 13: 978-1-138-55888-5 (pbk)
ISBN 13: 978-0-203-71340-2 (ebk)

Visit the Taylor & Francis Web site at http://www.taylorandfrancis.com and the CRC Press Web site at http://www.crcpress.com

To my parents Sankar K. Pal

To my beloved wife Julia Paul P. Wang

Contents

Preface

Pattern recognition and machine learning form a major area of research and development activity that encompasses the processing of pictorial and other nonnumerical information obtained from interactions among science, technology, and society. A motivation for this spurt of activity in this field is the need for the people to communicate with computing machines in their natural mode of communication. Another important motivation is that the scientists are concerned with the idea of designing and making intelligent machines that can carry out certain tasks as we human beings do. The most salient outcome of these is the concept of future-generation computing systems.

The ability to recognize a pattern is the first requirement for any intelligent machine. Pattern recognition is a must component of the so-called "Intelligent Control Systems" which involve processing and fusion of data from different sensors and transducers. It is also a necessary function providing "failure detection","verification", and "diagnosis task". Machine recognition of patterns can be viewed as a twofold task, consisting of learning the invariant and common properties of a set of samples characterizing a class and of deciding that a new sample is a possible member of the class by noting that it has properties common to those of the set of samples. Therefore, the task of pattern recognition by a computer can be described as a transformation from the measurement space to the feature space and finally to the decision space.

Various approaches so far proposed and experimented with for the design of pattern recognition systems can be broadly categorized into decision theoretic approach (both deterministic and probabilistic), syntactic approach, and connectionist approach. Methods and technologies developed under these categories may again be fuzzy set theoretic in order to handle uncertainties, arising from vague, incomplete, linguistic, overlapping patterns etc., at various stages of pattern recognition systems. One may note that the methods developed for pattern recognition and image processing are usually problem dependent. Moreover, many tasks involved in the process of recognizing a pattern need appropriate parameter selection and efficient search in complex spaces in order to obtain optimal solutions. This makes the process not only computationally intensive, but also leads to a possibility of losing the exact solution.

Genetic algorithms are randomized search and optimization techniques guided by the principles of evolution and natural genetics. They are efficient, adaptive, and robust search processes, producing near-optimal solutions and having a large amount of implicit parallelism. Therefore, the application of genetic algorithms for solving certain problems of pattern recognition (which need optimization of computation requirements, and robust, fast, and close approximate solution) appears to be appropriate and natural. The research articles on the integration of genetic algorithms and pattern recognition have started to come out.

The present book provides a collection of some new material, written by leading experts all over the world, demonstrating the various ways this integration can be made in order to develop intelligent recognition systems. This collection ranges from chapters providing insight into the theory of genetic algorithms, developing the pattern recognition theory in the light of genetic algorithms, to some of its applications in the areas of artificial neural networks and fuzzy logic.

The chapter by De, Ghosh, and Pal first of all provides, in condensed format, the basic principles and features of genetic algorithms. The necessity of incorporating the effects of parents' and grandparents' fitness value in computing the fitness of an individual string is then explained. This is supported by both the schema theorem and goodness of solutions. Procedures for automatic selection of the weighting coefficients representing the ancestors' contribution are also mentioned. The chapter by Vose and Wright presents a mathematical formulation of simple genetic algorithms in terms of Walsh transform (which is used in image processing and pattern recognition problems). Various theoretical results demonstrating the connection between them are provided. The third chapter written by Patnaik and Srinivas starts with a brief review on the recent work on adaptive strategies for modifying the control parameters of genetic algorithms. This is followed by a strategy of adapting the crossover and mutation probabilities based on the need for preserving good solutions. The chapter by Mathias, Whitley, Kusuma, and Stork deals with a comparative study of various genetic algorithms against a mutation-driven stochastic hill-climbing algorithm for optimization of noisy objective functions. Several hybrid genetic algorithms have also been developed with an application on geophysical static corrections of noisy seismic data.

The next two chapters provide some results of investigation enriching the theory of pattern classification and learning in the framework of genetic algorithms. An approach to develop general heuristics, for solving problems in knowledge-lean environments using genetic algorithms, is developed in the chapter by Wah. The generalization is attempted over the problem space which is not seen during learning. The chapter by Murthy, Bandyopadhyay, and Pal proves that as the number of training patterns tends to infinity, the performance of a genetic algorithm-based classifier, developed by the authors, approaches that of the Bayes' classifier. This finding is empirically verified on a data set with triangular distribution.

The seventh and eighth chapters are related to image processing. A modified scheme of chromosome encodings is suggested in the chapter by Van Hove and Verschoren, which utilizes trees and two-dimensional bit maps as its structure. This is useful to image processing problems where ordinary genetic algorithms, using linear string, do not function well. The chapter by Buckles, Petry, Prabhu and Lybanon, on the other hand, demonstrates that an application of genetic algorithms for automatic scene labeling of remote sensing imagery can really be beneficial. It is worthy to note that the heuristic knowledge about the domain constraints is represented by a semantic net.

The next two chapters are concerned with the application of genetic algorithms to artificial neural networks. The chapter by Romaniuk describes an approach for the automatic construction of neural networks for pattern classification. Genetic algorithms are applied here to locally train the network features using the perceptron rules. An attempt is also made to use this algorithm as a trans-dimensional learner which can automatically adjust the learning bias inherent to all learning systems. In the chapter by Gaudet, some new operators are introduced for the construction of logic-based neural networks where the response of the neurons is characterized by fuzzy set theory.

The ideal of "fuzzy decision making using genetic algorithms" forms the area of research contribution of the remaining three chapters. A classifier system with linguistic if–then rules is designed in the chapter by Ishibuchi, Murata, and Tanaka, where genetic algorithms are used to optimize the number of rules and to learn the grade of certainty associated with each rule. In the next chapter genetic algorithms are utilized by Janikow in optimizing the fuzzy knowledge components of a fuzzy decision tree both prior to and during its construction. This method also finds application in optimizing fuzzy rule-based systems. The last chapter, by Cooper and Vidal, describes a method of automatic generation of fuzzy controller by introducing a new encoding scheme for the rules which results in a more compact rule base.

This comprehensive collection provides a cross-sectional view of the research work that is currently being carried out applying the theory of genetic algorithms to different facets of pattern recognition systems, and makes the book unique of its own kind. The book may be used either in a graduate level as a part in the subject of pattern recognition and artificial intelligence, or as a reference book for the research workers in the area of genetic algorithms and its application to pattern recognition and machine-learning problems.

We take this opportunity to thank all the contributors for agreeing to write for this book. We owe a vote of thanks to Mr. Bob Stern of CRC Press, for his initiative and encouragement. The assistance provided by Ms. S. Bandyopadhyay, Mr. S. Das, Mr. I. Dutta, Mrs. Carol Nolte, and Mrs. Shirley Tsung Wang during the preparation of the book is also gratefully acknowledged. This work was carried out when Prof. S. K. Pal held a Jawaharlal Nehru Fellowship.

January 1996.   Sankar K. Pal
Paul P. Wang

Editors

Sankar K. Pal
Machine Intelligence Unit
Indian Statistical Institute
Calcutta, India

Paul P. Wang
Department of Electrical and Computer Engineering
Duke University
Durham, North Carolina, U.S.A.

Contributors

Sanghamitra Bandyopadhyay
Machine Intelligence Unit
Indian Statistical Institute
Calcutta, India

Bill P. Buckles
Center for Intelligent and Knowledge-based Systems
Department of Computer Science
Tulane University
New Orleans, Louisiana, U.S.A.

Mark G. Cooper
Department of Computer Science
University of California
Los Angeles, California, U.S.A.

Susmita De
Machine Intelligence Unit
Indian Statistical Institute
Calcutta, India

Vincent Charles Gaudet
Department of Electrical and Computer Engineering
University of Toronto
Toronto, Ontario, Canada

Ashish Ghosh
Machine Intelligence Unit
Indian Statistical Institute
Calcutta, India

Arthur Ieumwananonthachai
Coordinated Science Lab
University of Illinois
Center for Reliable and High-Performance Computing
Urbana-Champaign
Urbana, Illinois, U.S.A.

Hisao Ishibuchi
Department of Industrial Engineering
Osaka Prefecture University
Osaka, Japan

Cezary Z. Janikow
Department of Mathematics and Computer Science
University of Missouri
St. Louis, Missouri, U.S.A.

Anthony Kusuma
Advance Geophysical Corporation
Englewood, Colorado, U.S.A.

Yong-Cheng Li
Computer Science Department
University of Illinois
Urbana-Champaign
Urbana, Illinois, U.S.A.

Matthew Lybanon
Remote Sensing Applications Branch
Naval Research Laboratory
Stennis Space Center, Mississippi, U.S.A.

Srinivas Mandavilli
Motorola India Electronics Limited
Bangalore, India

Keith Mathias
Philips Laboratories
Briarcliff Manor, New York, U.S.A.

Tadahiko Murata
Department of Industrial Engineering
Osaka Prefecture University
Osaka, Japan

Chivukula A. Murthy
Machine Intelligence Unit
Indian Statistical Institute
Calcutta, India

Lalit M. Patnaik
Microprocessor Applications Lab
Indian Institute of Science
Bangalore, India

Frederick E. Petry
Center for Intelligent and Knowledge-based Systems
Department of Computer Science
Tulane University
New Orleans, Louisiana, U.S.A.

Devaraya Prabhu
Center for Intelligent and Knowledge-based Systems
Department of Computer Science
Tulane University
New Orleans, Louisiana, U.S.A.

Steve G. Romaniuk
Universal Problem Solvers, Inc.
Clearwater, Florida, U.S.A.

Christof Stork
Advance Geophysical Corporation
Englewood, Colorado, U.S.A.

Hideo Tanaka
Department of Industrial Engineering
Osaka Prefecture University
Osaka, Japan

Hugo Van Hove
Department of Mathematics and Computer Science
University of Antwerp, RUCA
Antwerp, Belgium

Alain Verschoren
Department of Mathematics and Computer Science
University of Antwerp, RUCA
Antwerp, Belgium

Jacques J. Vidal
Department of Computer Science
University of California
Los Angeles, California, U.S.A.

Michael D. Vose
Computer Science Department
The University of Tennessee
Knoxville, Tennessee, U.S.A.

Benjamin W. Wah
Coordinated Science Lab
University of Illinois
Center for Reliable and High Performance Computing
Urbana-Champaign
Urbana, Illinois, U.S.A.

Darrell Whitley
Department of Computer Science
Colorado State University
Fort Collins, Colorado, U.S.A.

Alden H. Wright
Computer Science Department
The University of Montana
Missoula, Montana, U.S.A.

GENETIC
ALGORITHMS
for
PATTERN
RECOGNITION

1

Fitness Evaluation in Genetic Algorithms with Ancestors' Influence

Susmita De, Ashish Ghosh, and Sankar K. Pal

Abstract   A new fitness evaluation criterion for Genetic Algorithms (GAs) is introduced where the fitness value of an individual is determined by considering its own fitness as well as that of its ancestors. The guidelines for selecting the weighting coefficients, both heuristically and automatically, which quantify the importance to be given on the fitness of the individual and its ancestors, are provided. The Schema theorem for the proposed concept is derived. The effectiveness of this new method has been demonstrated on the problems of optimizing complex functions. Results are found to be superior to those of the conventional genetic algorithms, both in terms of goodness of solution and the lower bound of the number of instances of good schemata.

1.1   Introduction

Genetic Algorithms (GAs) [1]–[3] are adaptive and robust computational procedures modeled on the mechanics of natural genetic systems. GAs act as a biological metaphor and try to emulate some of the processes observed in natural evolution. They are viewed as randomized, yet structured, search and optimization techniques. They express their ability by efficiently exploiting the historical information to speculate on new offspring with expected improved performance [1]. GAs are executed iteratively on a set of coded solutions, called a *population*, with three basic operators: *selection/reproduction, crossover*, and *mutation*.

In order to find out the optimum solution of a problem, a GA starts from a set of assumed solutions (chromosomes) and evolves different but better sets (of solutions) over a sequence of iterations. In each generation (iteration) the objective function (fitness measuring criterion) determines the suitability of each solution and, based on these values, some of them (which are called parent chromosomes) are selected for reproduction. The number of copies reproduced by an individual parent is expected to be directly proportional to its fitness value, thereby embodying the natural selection procedure, to some extent. The procedure thus selects better (highly fitted) chromosomes and worse ones are eliminated. Hence, the performance of a GA depends on the fitness evaluation criterion, to a large extent. Genetic operators are applied on these (selected) parent chromosomes and new chromosomes (offspring) are generated. Conventional genetic algorithms (CGAs) consider only the fitness value of the chromosome under consideration for measuring its suitability for selection for the next generation, *i.e.*, the fitness of a chromosome is a function of the functional value of the objective function. Fitness of a chromosome $x = g[f(x)]$, where $f(x)$ is the objective function and g is another function which by operating on $f(x)$ gives the fitness value. Hence, a CGA does not discriminate between two identical offspring, one produced from better (highly fit) parents and the other from comparatively weaker (low fit) parents. In nature, normally an offspring is more fit (suitable) if its ancestors (parents) are better, *i.e.*, an offspring possesses some extra facility to exist in its environment if it belongs to a better family (ancestors are highly fitted). In other words, the fitness of an individual depends also on the fitness of its ancestors in addition to its own fitness. This is perhaps the basic intuition to give more weightage to highly fitted chromosomes (due to better ancestors) to produce more for the next generation.

This chapter is an attempt in this line by providing a new concept for measuring the fitness of an individual by considering its own fitness as well as that of its ancestors, *i.e.*, fitness of a chromosome $x = g[f(x), a_1, a_2, \ldots, a_n]$, where a_is are the fitness values of its ancestors. The function g may be of various types. The weightage given to the fitness of the ancestors can be assigned heuristically at the beginning and kept fixed throughout the procedure or it may be varying/adaptive. Heuristics are given for choosing the weighting factors manually, and a procedure for evolving them automatically has also been developed. The merit of the proposed algorithm as compared to that of the CGA is explained using the Schema theorem. The effectiveness of this concept has been demonstrated experimentally on the problems of complex function optimization. The performance is seen to be enhanced by the new fitness evaluation criterion.

The rest of this chapter is organized as follows: in Section 1.2 a brief introduction to genetic algorithms is provided; Section 1.3 describes the proposed fitness evaluation methodology; implementation details are given in Section 1.4; analysis of results is presented in Section 1.5; and concluding remarks are put in Section 1.6.

1.2 Genetic Algorithms: Basic Principles and Features

GAs (sometimes we will be referring to them as CGAs) are intended to mimic some of the processes observed in natural evolution in the following ways [2]:

- Evolution is a process that operates on encoding of biological entities, rather than on the living beings themselves, *i.e.,* evolution takes place at the level of chromosomes.
  Similarly, GAs operate on encoding of possible solutions (called chromosomes) of the problems, by manipulating strings of characters of an alphabet.

- Natural selection is the link between a chromosome and its performance (measured on its decoded version). Nature obeys the principle of Darwinian "survival of the fittest"; the chromosomes with high fitness values will, on the average, reproduce more often than those with low fitness values.
  In GAs, selection of chromosomes also tries to mimic this procedure. Highly fitted chromosomes will reproduce more often at the cost of lower fitted ones.

- Biological evolution has no memory. Here, nature acts as the environment and the biological entities are modified to adapt in their environment. Whatever nature knows about the evolution of good chromosomes is contained in the set of chromosomes possessed by the current individuals and in the chromosome decoding procedure.
  Likewise, GAs also operate on chromosomes blindly. They use only the objective function information which acts as the environment. Based on this information, each chromosome is evaluated and during the selection process more importance is given on choosing chromosomes having high fitness values.

- Like natural evolution, variation of the entities in GAs is introduced when reproduction occurs. Crossover and mutation are the basic operators for reproduction.

The GA-based evolution starts from a set of individuals (assumed solution set for the function to be optimized) and proceeds from generation to generation through genetic operations. Replacement of an old population with a new one is known as generation when the generational replacement technique (replace all the members of old population with the new ones) is used. Another reproduction technique, called steady-state reproduction [4, 5], replaces one or more individuals at a time instead of the whole population. As mentioned before, GAs require only a suitable objective function which acts as the environment in order to evaluate the suitability of the derived solutions (chromosomes). A schematic diagram of the basic structure of a genetic algorithm is shown in Figure 1.1.

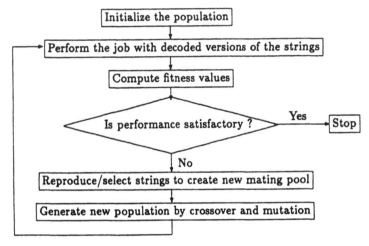

FIGURE 1.1
Basic steps of a genetic algorithm.

A GA typically consists of the following components:

- A population of strings or coded possible solutions (biologically referred to as chromosomes)

- A mechanism to encode a possible solution (mostly as a binary string)

- Objective function and associated fitness evaluation techniques

- Selection/reproduction procedure

- Genetic operators (crossover and mutation)

- Probabilities to perform genetic operations

Let us briefly describe these components:

Population—To solve an optimization problem, GAs start with the chromosome (structural) representation of a parameter set $\{x_1, x_2, \ldots, x_p\}$. The parameter set is to be coded as a finite length string over an alphabet of finite length. Usually, the chromosomes are strings of 0's and 1's. For example, let $\{a_1, a_2, \ldots, a_p\}$ be a realization of the parameter set and the binary representation of $a_1, a_2, \ldots, a_p$ be 10110, 00100, …, 11001, respectively. Then the string 10110 00100 … 11001 is a chromosome representation of the parameter set $\{a_1, a_2, \ldots, a_p\}$. It is evident that the number of different chromosomes (strings) is 2^l where, l is the string length. Each chromosome actually refers to a coded possible solution. A set of such chromosomes in a generation is called a population. The size of a population may vary from one generation to another or it can be constant. Usually, the initial population is chosen randomly.

Encoding/decoding mechanism—It is the mechanism to convert the parameter values of a possible solution into binary strings resulting into chromosome representation. If the solution of a problem depends on p parameters and if we want to encode each parameter with a binary string of length q, then the length of each chromosome will be $p \times q$. Decoding is just the reverse of encoding.

Objective function and associated fitness evaluation techniques—The fitness/objective function is chosen depending on the problem. It is chosen in a way such that highly fitted strings (possible solutions) have high fitness values. It is the only index to select a chromosome to reproduce for the next generation.

Selection/reproduction procedure—The selection/reproduction process copies individual strings (called parent chromosomes) into a tentative new population (known as mating pool) for genetic operations. The number of copies reproduced for the next generation by an individual is expected to be directly proportional to its fitness value, thereby mimicking the natural selection procedure to some extent. Roulette wheel parent selection [1] and linear selection [2] are the most frequently used selection procedures.

Genetic operators are applied on parent chromosomes and new chromosomes (called offspring) are generated. Frequently used genetic operators are described below.

Crossover—The main purpose of crossover is to exchange information between randomly selected parent chromosomes with the aim of not losing any important information (minimum disruption of the structures of the chromosomes that are selected for genetic operation). Actually, it recombines genetic material of two parent chromosomes to produce offspring for the next generation. The crossover operation can be viewed as a three-step process. In the first step, pairs of chromosomes (called mating pairs) are chosen from the mating pool. The second step determines, on the basis of crossover probability (by generating a random number in [0, 1] and checking whether this number is less than the crossover probability), whether these pairs of chromosomes should go for crossover or not. Interchange of chromosome segments between mating pairs is done in the third step. The number of segments and the length of each segment to be exchanged depend on the type of crossover technique. Some of the commonly used crossover techniques are one-point crossover [6], two-point crossover [7], multiple-point crossover [8], shuffle-exchange crossover [7], uniform crossover [5, 7] etc. To illustrate how the segments of two parent chromosomes are swapped, let us consider the one-point crossover technique. Here, a position k is selected uniformly at random between 1 and $l - 1$, where l is the string length (greater than 1). Two new strings are created by swapping all characters from position $(k + 1)$ to l. Let:

$$a = 11000\ 10101\ 01000 \ldots 01111\ 10001$$
$$b = 10001\ 01110\ 11101 \ldots 00110\ 10100$$

be two strings (parents) selected for the crossover operation and the generated random number be 11 (eleven). Then the newly produced offspring (swapping all characters after position 11) will be

$$a' = 11000\ 10101\ 01101 \ldots 00110\ 10100$$
$$b' = 10001\ 01110\ 11000 \ldots 01111\ 10001$$

Mutation—The main aim of mutation is to introduce genetic diversity into the population. Sometimes, it helps to regain the information lost in earlier generations. Like natural genetic systems, mutation in GAs is also made occasionally. A random position of a random string is selected and is replaced by another character from the alphabet. In the case of binary representation it negates the bit value and is known as bit mutation. For example, let the third bit of string a, shown above, be selected for mutation. Then the transformed string after mutation will be

$$11100\ 10101\ 01000 \ldots 01111\ 10001$$

Normally mutation rate is kept fixed. To sustain diversity (which may be lost due to crossover and very low mutation rate) into the population, Whitley et al. [9] proposed a technique called *adaptive mutation*, where the probability to perform mutation operation is made to increase (instead of keeping it fixed) with increase of genetic homogeneity in the population. Mutation is not always worth performing. High mutation rate can lead the genetic search to a random one. It may change the value of an important bit and thereby affect the fast convergence to a good solution. Moreover, it may slow down the process of convergence at the final stage of GAs. Recently Bhandari et al. [10] have proposed a new mutation technique known as *directed mutation*.

Probabilities to perform genetic operations—The probability to perform crossover operation is chosen in a way so that recombination of potential strings (highly fitted chromosomes) increases without any disruption. Generally, the crossover probability lies in-between 0.6 to 0.9 [1, 2].

Since mutation occurs occasionally, it is clear that the probability of performing mutation operation will be very low. Typically the value lies between 0.001 to 0.01 [1, 2].

Crossover and mutation probabilities may be kept fixed throughout the operation of a GA or they may also be made adaptive (determined automatically depending on the environment) [11] to improve the performance, if possible.♣

In standard GA (SGA), we do not preserve the best possible solution obtained so far, thereby increasing the chance of losing the obtainable best possible solution.

Elitist strategy overcomes this problem by copying the best member of each generation into the next one. Though this strategy may increase the speed of dominance of a population by a potential string (string with high fitness value), it enhances the performance of a GA using generational replacement. The concept of distributed GAs [9] and parallel GAs [12] has also been introduced recently. ♣

Distinguishing characteristics—The following features facilitate GAs to enhance their applicability over most of the commonly used optimization and searching strategies:

- GAs work simultaneously on multiple points and not on a single point, which helps to introduce a large amount of implicit parallelism in the computational procedure. This feature also prevents a GA from getting stuck at local optimal, to some extent.

- GAs work with the coded parameter set, not with the parameters themselves. Hence, the resolution of the possible search space can be controlled by varying the coding length of the parameters.

- In GAs, the search space need not be continuous (unlike a calculus-based search method).

- GAs are blind. They use only the objective function information and do not need any auxiliary information, like derivative of the optimizing function.

- GAs use probabilistic transition rules, not deterministic ones.

Recently, GAs are finding widespread applications in solving problems, requiring efficient and effective searches, in business, scientific, and engineering circles [2, 13, 14] e.g., synthesis of neural network architectures, traveling salesman problem, graph coloring, scheduling, numerical optimization, classifier systems, pattern recognition.

1.3 A New Fitness Evaluation Criterion

In this section we describe a new method for evaluating the fitness of an individual chromosome by considering the effect of fitness of its ancestors along with its own fitness, for its selection for reproduction. As mentioned before, in CGA, chromosomes are selected for reproduction based on their own fitness values. It does not consider any influence of its ancestors (predecessors). However, in nature, "family background" plays a significant role to determine the characteristics and suitability of offspring; offspring from better families (highly fitted ancestors) invariably possess some extra advantages to be treated as fit in an environment.

This natural phenomenon motivates us to consider the effect of fitness of ancestors (or parents) along with the fitness of the individuals to measure their fitness for survival. Here, good features are made to propagate to the subsequent generations by the introduction of the effect of ancestors (parents) to determine the fitness of the offspring.

If we want to consider the influence of n ancestors while measuring the fitness of an individual, the modified fitness value (MFV) of an individual x will be

$$MFV = g(fit, a_1, a_2, \ldots, a_n), \tag{1.1}$$

where fit is the original fitness of the individual x, and $a_1, a_2, \ldots, a_n$ are the fitness of its n ancestors. A simple form of g may be as follows:

$$MFV = \alpha \times fit + \sum_{i=1}^{n} \beta_i \times a_i, \tag{1.2}$$

where α and β_is are the weighting factors quantifying the effect of fitness of the individual under consideration and those of its ancestors. For convenience, we have taken $\alpha + \sum_{i=1}^{n} \beta_i = 1$. These weighting coefficients may be taken as static (initially set to some value, based on heuristics, and kept constant throughout the procedure) or dynamic (these coefficients will change automatically with the environment). Automatic evolution (depending on the environment) of α and β_is sounds more logical.

Now, if we want to consider the effect of the previous generation only, we need to take into account the fitness of the parents (say, p_1 and p_2) only; in that case Equation 1.2 reduces to

$$MFV = \alpha \times fit + \beta_1 \times p_1 + \beta_2 \times p_2. \tag{1.3}$$

As a special case, if we choose $\beta_1 = \beta_2 = \beta$ (*i.e.*, we put equal weightage to both parents), then the modified fitness value will be

$$MFV = \alpha \times fit + \beta \times (p_1 + p_2). \tag{1.4}$$

If we go two generations ahead, fitness of parents (p_1 and p_2) as well as that of grandparents (say, gp_1, gp_2, gp_3 and gp_4) will come into consideration. In this case, the number of ancestors to be considered will be six (two parents and four grandparents), *i.e.*, $n = 6$, and Equation 1.2 will take the form:

$$MFV = \alpha \times fit + \sum_{i=1}^{2} \beta_i p_i + \sum_{j=1}^{4} \gamma_j gp_j, \tag{1.5}$$

where β_is and γ_js are the weightage of parents and grandparents, respectively. As a particular case, where all βs are equal and all γs are equal, this equation reduces to

$$MFV = \alpha \times fit + \beta \times (p_1 + p_2) + \gamma \times (gp_1 + gp_2 + gp_3 + gp_4). \quad (1.6)$$

It is intuitively clear that $\beta_i > \gamma_j$, $\forall i$ and $\forall j$ and it has also been found experimentally that the influence of grandparents is less compared to the influence of parents for survival of an individual. Hence for all practical purposes, we restrict ourselves going beyond two generations in order to consider the effect of "family background".

In this context it is to be noted that the influence of grandparents and their ancestors comes into account implicitly if we consider the influence of parents in order to determine the suitability of an individual. For measuring the fitness of chromosomes at kth generation, the fitness of chromosomes of $(k-1)$th generation is considered, since they are the parents of chromosomes of kth generation. Similarly, for getting the fitness of these parents (chromosomes of $[k-1]$th generation) the fitness values of chromosomes of $(k-2)$th generation were considered. Thus, eventually, the effect of fitness of chromosomes of $(k-2)$th generation (grandparents of the chromosomes of kth generation) reaches the chromosomes of the kth generation.

1.3.1 Selection of Weighting Coefficients (α, β, γ)

The values of weighting coefficients (*i.e.*, $\alpha, \beta_i, \gamma_j$) may be assigned heuristically at the beginning of the procedure and kept constant throughout it, or they may be assigned randomly at the onset and evolve automatically depending on the environment. For heuristic assignment, one may consider $\alpha > \beta_i, \gamma_j$, *i.e.*, more weightage is given to the offspring itself than to its ancestors. Further, their values may be set depending on whether mutation has occurred on a chromosome or not (as mutation normally changes the characteristics of the individuals drastically). If it occurs, the value of α is considered to be greater than that in the previous case. This choice may be justified as follows. Mutation is supposed to bring very high change in a chromosome. Thus if mutation occurs on a chromosome and if the chromosome becomes comparatively highly fitted for the next generation, then the effect of parents may be comparatively less than if mutation does not occur. Since the chromosome is much better with its own fitness, it does not need more support from its parents. On the other hand, if the chromosome becomes very bad after mutation, then, also, no support can make it fit to be selected for the next generation. Hence as a whole, less weightage will be given to the fitness of parents and more weightage to the individual chromosomes, in case mutation occurs.

To evolve the weighting coefficients automatically (adaptively) over sequence of generations, the weighting factors are taken as parameters of the problem. Hence, some fields are kept for α, β_i, γ_j in the chromosome representation of the possible solutions. Due to crossover and mutation, values of α, β_i, γ_j will be changing with time and thus they will be evolved automatically. As we choose better chromosomes from generation to generation, the evolved values of α, β_i, and γ_j (which are parts of the chromosomes) will be more suitable for that environment.

1.3.2   The Schema Theorem and the Influence of Parents on the Offspring

The Schema theorem [1] estimates the lower bound of the number of instances of different schemata at any point of time. According to this theorem, a short-length, low-order, above-average schema will receive exponentially increasing instances in subsequent generations at the expense of below average ones. In this section we derive the Schema theorem for the proposed concept (let us call it "modified genetic algorithms" [$MGAs$]) and find the lower bound of the number of instances of a schema. We also compare this bound with that of the CGA. The notations that we will be using for this purpose is explained below:

x:	average fitness value of the population at generation t (stage a, say)
y:	average fitness value of the population at mating pool (after selection of chromosomes from generation t) (stage b, say)
z:	average fitness value of the population at generation $(t + 1)$ (stage c, say)
w:	average fitness value of the population at mating pool (after selection of chromosomes from generation $[t + 1]$) (stage d, say)
u:	average fitness value of the population at generation $(t + 2)$ (stage e, say)
h:	a short-length, low-order, above average schema
$\overline{f^r}$:	average fitness value of the population at stage r for the CGA
$\overline{f}^{*r}$:	average fitness value of the population at stage r for the MGA
$\overline{f_h}^r$:	average fitness value of schema h at stage r for the CGA
$\overline{f_h}^{*r}$:	average fitness value of schema h at stage r for the MGA
$m(h, t)$:	m instances of a schema h in a population at generation t for the CGA
$m^*(h, t)$:	m instances of a schema h in a population at generation t for the MGA
$\delta(h)$:	the defining length of schema h
$o(h)$:	order of schema h
L:	length of a chromosome

We now derive the expression for the expected number of instances of schema h, *i.e.*, determine $m(h, t + 2)$ from $m(h, t + 1)$ using both the CGA and the MGA (considering the influence of parents only).

For the CGA,

$$m(h, t + 2) \geq m(h, t + 1) \times \frac{\overline{f_h}^c}{\overline{f}^c} \times \{1 - p_c \times \frac{\delta(h)}{L - 1} - o(h)p_m\}. \quad (1.7)$$

In the MGA, the *MFV* of a chromosome is obtained using Equation 1.4 (if we consider equal influence of both the parents). Accordingly, the average fitness value of the population (at stage c) gets changed. Let the modified average fitness value be $\overline{f}^{*c}$ and this, in turn, changes the average fitness value of the schema. The fitness values of the parent chromosomes of stage c are obtained from stage b. Hence, the modified average fitness value of the schema h ($\overline{f_h}^{*c}$) can be represented as

$$\overline{f_h}^{*c} = g(\overline{f_h}^c, \overline{f_h}^b), \quad (1.8)$$

where g is a function. The chromosomes from generation $(t + 1)$ (*i.e.*, stage c) are selected based on their modified fitness values and modified average fitness value $(\overline{f}^{*c})$. Hence, the order of selection of chromosomes may be different from that of the CGA.

So, the expected number of instances of the schema h at generation $(t+2)$ (stage e) for the MGA will be

$$m^*(h, t + 2) \geq m(h, t + 1) \times \frac{g(\overline{f_h}^c, \overline{f_h}^b)}{\overline{f}^{*c}} \times \{1 - p_c \times \frac{\delta(h)}{L - 1} - o(h)p_m\}. \quad (1.9)$$

Let $\overline{f}^c = \theta \times \overline{f}^{*c}$. Substituting the value of $\overline{f}^{*c}$ in Equation 1.9, we get,

$$m^*(h, t+2) \geq m(h, t + 1) \times \frac{\theta \, g(\overline{f_h}^c, \overline{f_h}^b)}{\overline{f}^c} \times \{1 - p_c \times \frac{\delta(h)}{L - 1} - o(h)p_m\}. \quad (1.10)$$

Let us denote the right-hand side of Equations 1.7 and 1.10 by A and B, respectively. Thus, if $\overline{f_h}^c < \theta \, g(\overline{f_h}^c, \overline{f_h}^b)$, then $A < B$, i.e., the lower bound of $m^*(h, t + 2)$ will be more than that of $m(h, t + 2)$. In this context it may be noted that for linear normalization selection procedure, $\theta = 1.\spadesuit$

To illustrate this further, let us consider the situation corresponding to Equation 1.4, where

$$g(\overline{f_h}^c, \overline{f_h}^b) = \alpha \overline{f_h}^c + \beta \overline{f_h}^b, \quad (1.11)$$

with $\alpha + \beta = 1$.

Case 1: If $\overline{f_h}^b = \overline{f_h}^c$, then

$$g(\overline{f_h}^c, \overline{f_h}^b) = \alpha \overline{f_h}^c + \beta \overline{f_h}^c$$

$$= (\alpha + \beta)\overline{f_h}^c$$

$$= \overline{f_h}^c$$

Thus we get $B = A\theta$. Hence, $B > A$, if $\theta > 1$.

Case 2: If $\overline{f_h}^b > \overline{f_h}^c$, then

$$g(\overline{f_h}^c, \overline{f_h}^b) = \alpha\overline{f_h}^c + \beta(\overline{f_h}^c + k), \qquad k > 0$$

$$= \alpha\overline{f_h}^c + \beta\overline{f_h}^c + \beta k$$

$$= (\alpha + \beta)\overline{f_h}^c + \beta k$$

$$= \overline{f_h}^c + \beta k$$

$$> \overline{f_h}^c$$

Thus from this derivation we obtain $B > A\theta$. Hence, $B > A$, if $\theta \geq 1$.

Case 3: If $\overline{f_h}^b < \overline{f_h}^c$, then

$$g(\overline{f_h}^c, \overline{f_h}^b) = \alpha\overline{f_h}^c + \beta(\overline{f_h}^c - k), \qquad k > 0$$

$$= \alpha\overline{f_h}^c + \beta\overline{f_h}^c - \beta k$$

$$= (\alpha + \beta)\overline{f_h}^c - \beta k$$

$$= \overline{f_h}^c - \beta k$$

Now, one of the following three conditions will arise:

- $\theta(\overline{f_h}^c - \beta k) > \overline{f_h}^c$, then $B > A$
- $\theta(\overline{f_h}^c - \beta k) = \overline{f_h}^c$, then $B = A$
- $\theta(\overline{f_h}^c - \beta k) < \overline{f_h}^c$, then $B < A$

Thus we can conclude from the above illustration that if $\overline{f_h}^b \geq \overline{f_h}^c$ and $\theta > 1$, then $B > A$, i.e., the lower bound of $m^*(h, t+2)$ will be more than that of $m(h, t+2)$. Otherwise, the performance will depend on the values of β and θ.

1.4   Implementation

The effectiveness of the aforesaid concept has been demonstrated on the problems of optimizing complex functions. We have used three different functions f1, f2, and f3 (described in Table 1.1). The first two are univariate functions, whereas the third one is a bivariate one (which is first described by Schaffer [15] and later by Davis [2]).

Table 1.1   Functions Used for Optimization

Function	Functional form	Domain	Maximum value
f_1	$2 + e^{x-10}\cos(x-10)$, if $x \leq 10.0$ $2 + e^{10-x}\cos(x-10)$, if $x > 10.0$	[0, 20]	3.00
f_2	$\displaystyle\sum_{i=1}^{10}(x-2i)$	[0, 20]	3.72×10^9
f_3	$0.5 - \dfrac{\{\sin(\sqrt{(x^2+y^2)})\}^2 - 0.5}{\{1.0+0.001(x^2+y^2)\}^2}$	[−100, 100]	1.0

The complex behavior of the functions is depicted in graphical form in Figures 1.2 to 1.4. f1 has only one maximum. f2 has ten maxima with a global maximum at $x = 0$. f3 is a rapidly varying multimodal function with several close oscillating hills and valleys with a global maximum at $x = y = 0$.

1.4.1   Selection of Genetic Parameters

To optimize these functions, the following steps are adopted for the genetic algorithms:

- Binary representation is used for chromosomes, *i.e.*, a chromosome is a bit string.

- The substring length of each parameter (variable) has been taken as 22. Values of the parameters α, β, and γ are encoded by 10 bits.

- Population size is kept fixed at 20 throughout the simulation. Generational replacement technique has been used.

- The objective function is the identity function, *i.e.*, fitness of a chromosome is equal to its functional value. Higher the functional value, better is the chromosome. Both the elitist model (by copying the best member of each generation into the next one to replace the lowest fitted string) and the standard GA (SGA, *i.e.*, without elitism) are implemented.

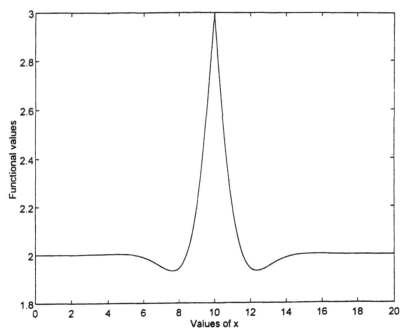

FIGURE 1.2
Sketch of the function

$2 + e^{x-10} \cos(x - 10)$, if $x \leq 10.0$

$2 + e^{x-10} \cos(x - 10)$, if $x > 10.0$

- Linear normalization selection procedure (which works better in a close competitive environment) is adopted. The difference between successive fitness values has been taken as 1, and the minimum fitness value has been kept to 1.

- The number of copies produced by the ith individual (chromosome) with normalized fitness value f_i in a population of size n is taken as round(c_i), where

$$c_i = \frac{n \times f_i}{\displaystyle\sum_{i=1}^{n} f_i}$$

and round(c_i) gives the nearest integer of the real number c_i. If round(c_i) $\leq$ 1, then the number of copies reproduced is taken as 1, provided the size of the population is not exceeded.

- The crossover and mutation probabilities are taken as 0.8 and 0.008, respectively. One point crossover operation is performed.

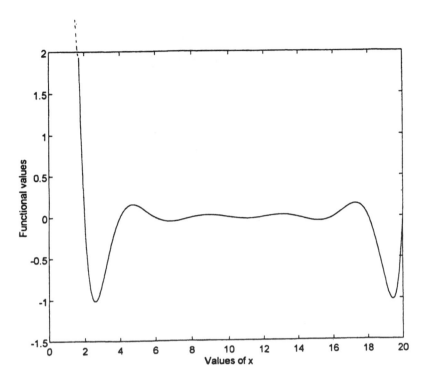

FIGURE 1.3

Sketch of the function $\displaystyle\sum_{i=1}^{10}(x - 2i)$.

1.4.2   Various Schemes

The investigation has been made through several schemes (corresponding to different values of α, β_i, γ_j) to find out the utility of the proposed concept. A few of them are described below for typical illustration.

Scheme 1: Here the effect of only parents are considered; we choose $\alpha = 0.5$ and $\beta_1 = \beta_2 = \beta = 0.25$ (Equation 1.4).

Scheme 2: Similar to *Scheme 1*, but the weightage is given depending on whether mutation has occurred or not in a chromosome. If mutation occurs on a chromosome, $\alpha = 0.8$ and $\beta_1 = \beta_2 = 0.1$; otherwise $\alpha = 0.5$ and $\beta_1 = \beta_2 = 0.25$.

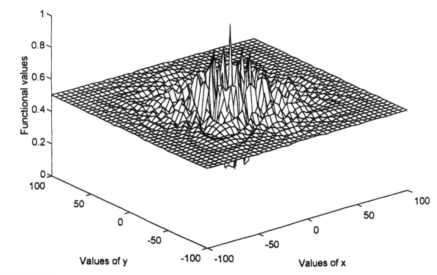

FIGURE 1.4

Sketch of the function $0.5 - \dfrac{\{\sin(\sqrt{(x^2 + y^2)})\}^2 - 0.5}{\{1.0 + 0.001(x^2 + y^2)\}^2}$.

Scheme 3: Similar to *Scheme 1*, but the weighting coefficients are evolved automatically. We use Equation 1.4. Since $\alpha + \beta = 1$, we need to evolve only one weighting factor (say, α).

Scheme 4: Similar to *Scheme 3*, but the amount of importance given to different parents is different (Equation 1.3). We restrict β_1 and β_2 in [0,0.5). Since, $\alpha + \beta_1 + \beta_2 = 1$, it ensures that some weightage will always be there for the individual chromosome, *i.e.*, $\alpha > 0$.

Scheme 5: The influence of both parents and grandparents is taken into account with the above mentioned schemes using Equations 1.5 and 1.6.

The algorithm has been run for 1000 iterations in each simulation. Five simulations were performed. Same seed value has been taken for all the schemes. Seed values for different simulations are different. The mean (of all the simulations) of the average (over the whole population) fitness values and the mean (of all the simulations) of the maximum fitness values (of the best chromosome in each generation) have been considered as performance measures.

1.5   Analysis of Results

Figures 1.5 and 1.6 display the performance curves (variation of average fitness value with generation using the nonelitist version) of the different GA-based schemes corresponding to the functions f2 and f3. Table 1.2 indicates the maximum fitness values of different schemes. It is evident from the table and the performance curves that a "better" family facilitates its children to be suitable in its environment. The results show that the performance of the proposed algorithm is better than that of the CGA in most of the cases for the functions f2 and f3 (which are relatively complex ones). For a simple function like f1, it did not produce any improvement.

Table 1.2   Maximum Fitness Values (Averaged over Five Simulations)

Scheme	f1		f2		f3	
no.	SGA	Elitism	SGA	Elitism	SGA	Elitism
CGA	2.999996	2.999995	2.23×10^9	2.23×10^9	0.945580	0.956583
1	2.999995	2.999996	2.97×10^9	2.23×10^9	0.951081	0.984782
2	2.999995	2.999996	2.97×10^9	2.23×10^9	0.976589	0.931675
3	2.999997	2.999997	2.97×10^9	2.97×10^9	0.922882	0.990261
4	2.999996	2.999996	2.97×10^9	2.97×10^9	0.979280	0.984782

Table 1.3   Evoluted α of the Best Chromosome for Different Simulations (*Scheme 3*) Using SGA

Simulation no.	f1	f2	f3
1	0.2227	0.1133	0.2559
2	0.6016	0.9824	0.0723
3	0.3672	0.9512	0.8398
4	0.7051	0.7822	0.9199
5	0.1006	0.3350	0.3643

In the case of automatic evolution of weighting factors (Figures 1.5b, 1.5c and 1.6b, 1.6c), the system needs some more time to adapt with the environment in order to generate improved results as compared to the CGA. Unequal weightage ($\beta_1 \neq \beta_2$) performs better than equal weightage, *i.e.*, $\beta_1 = \beta_2$ (Figures 1.5c and 1.6c). Further, improvement in performance under the elitist model is not as high as in the nonelitist version, *i.e.*, SGA (Table 1.2).

As an illustration, automatically evolved weighting factors of the best chromosomes (of last generation) of each simulation are presented in Table 1.3 (for

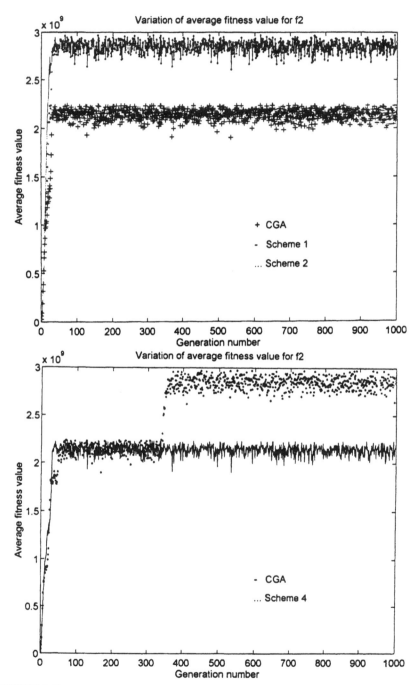

FIGURE 1.5
(a and b) Variation of average fitness value with generation using the nonelitist version of different GA-based schemes corresponding to function f2.

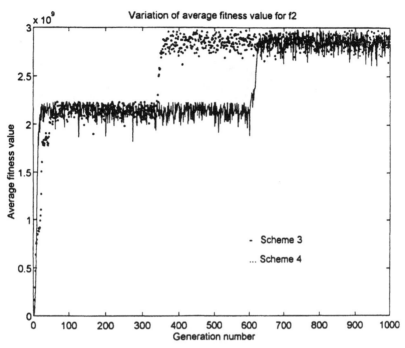

FIGURE 1.5
(c) Variation of average fitness value with generation using the nonelitist version of different GA-based schemes corresponding to function f2.

Table 1.4 Evoluted β_1 and β_2 of the Best Chromosome for Different Simulations (*Scheme 4*) Using SGA

	f1		f2		f3	
Simulation no.	β_1	β_2	β_1	β_2	β_1	β_2
1	0.3545	0.0005	0.1270	0.0010	0.0508	0.1255
2	0.2002	0.3882	0.0518	0.0166	0.0176	0.0625
3	0.1006	0.2207	0.1592	0.2954	0.0483	0.0903
4	0.0132	0.0303	0.1480	0.0435	0.1372	0.1860
5	0.0068	0.3633	0.4761	0.0230	0.1387	0.0317

Scheme 3) and Table 1.4 (for *Scheme 4*) for all the functions. These values may be compared with those considered in *Schemes 1* and 2.

Results under *Scheme 5* are not mentioned here because the influence of grandparents is seen to be not very effective. Study has also been conducted with different values of mutation and crossover probability. Results are found to be similar. *Scheme 1* with various combinations of α and β also resulted in similar performance.

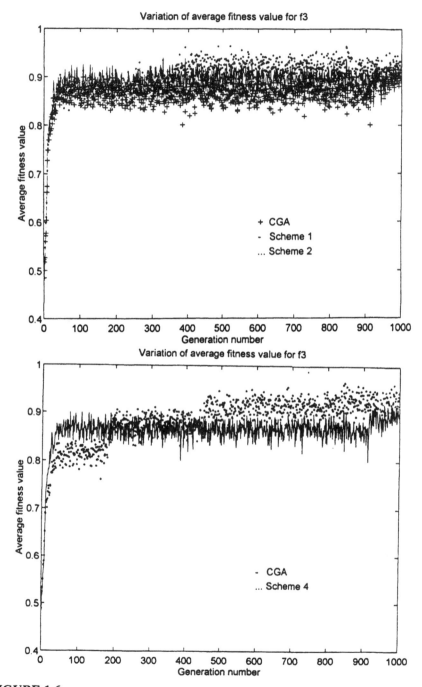

FIGURE 1.6
(a and b) Variation of average fitness value with generation using the nonelitist version of different GA-based schemes corresponding to function f3.

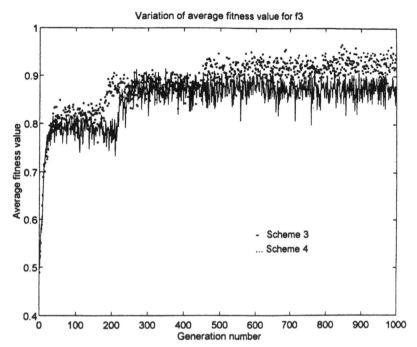

FIGURE 1.6
(c) Variation of average fitness value with generation using the nonelitist version of different GA-based schemes corresponding to function f3.

1.6 Conclusions

A new fitness evaluation criterion has been introduced for GAs which considers the effect of fitness of ancestors (predecessors) in addition to the fitness of the individual itself. Selection of chromosomes is made based on these modified fitness values. The Schema theorem for this new model is derived. Some conditions are found where the proposed concept leads to superior performance compared to the CGA in terms of the lower bound of the number of instances of a good schema in subsequent generations. The effectiveness of the proposed concept has been demonstrated by optimizing complex multimodal functions. Different schemes are provided to select the amount of weightage to be given to the ancestors. A method is also suggested to determine the weighting coefficients automatically. It has been found experimentally that the proposed concept outperforms the CGA for complex optimization problems having a large number of local optima in the search space. The performance is seen to be more appreciable for the nonelitist version than its elitist counterpart.

Acknowledgments

Ms. Susmita De is grateful to the Department of Atomic Energy, Government of India, for providing a research fellowship. This work was done when Prof. S. K. Pal held Jawaharlal Nehru Fellowship.

References

[1] Goldberg, D.E., *Genetic Algorithms in Search, Optimization and Machine Learning*, Addison-Wesley, Reading, MA, 1989.

[2] Davis, L., Ed., *Handbook of Genetic Algorithms*, Van Nostrand Reinhold, New York, 1991.

[3] Michalewicz, Z., *Genetic Algorithms + Data Structure = Evolution Programs*, Springer-Verlag, Berlin, 1992.

[4] Whitley, D. and Kauth, J., Genitor: a different genetic algorithm, in *Proc. of the Rocky Mountain Conf. on Artificial Intelligence*, Denver, CO, 1988, 118.

[5] Syswerda, G., Uniform crossover in genetic algorithms, in *Proc. 3rd Int. Conf. on Genetic Algorithms*, Morgan Kaufmann, San Mateo, CA, 1989, 2.

[6] Holland, J.H., *Adaptation in Natural and Artificial Systems*, University of Michigan Press, Ann Arbor, 1975.

[7] Eshelman, L.J., Caruana, R.A., and Schaffer, J., Biases in the crossover landscape, in *Proc. 3rd Int. Conf. on Genetic Algorithms*, Morgan Kaufmann, San Mateo, CA, 1989, 10.

[8] Frantz, D.R., Non-linearities in Genetic Adaptive Search, *Ph.D. dissertation*, University of Michigan, Ann Arbor, 1972.

[9] Whitley, D., Starkweather, T., and Bogart, C., Genetic algorithms and neural networks: optimizing connections and connectivity, *Parallel Comput.*, 14, 347, 1990.

[10] Bhandari, D., Pal, N.R., and Pal, S.K., Directed mutation in genetic algorithms, *Inf. Sci.*, 79, 251, 1994.

[11] Srinivas, M. and Patnaik, L.M., Adaptive probabilities of crossover and mutation in genetic algorithms, *IEEE Trans. Syst. Man Cybern.*, 24(4), 656, 1994.

[12] Maniezzo, V., Genetic evolution of the topology and weight distribution of neural networks, *IEEE Trans. Neural Networks*, 5(1), 39, 1994.

[13] Gelsema, E.S., Ed., Special issue on genetic algorithms, *Pattern Recognition Letters*, 16(8), 1995.

[14] Forrest, S., Ed., *Proc. 5th Int. Conf. on Genetic Algorithms*, Morgan Kaufmann, San Mateo, CA, 1993.

[15] Schaffer, J.D., Caruana, R.A., Eshelman, L.J., and Das, R., A study of control parameters affecting online performance of genetic algorithms for function optimization, in *Proc. 3rd Int. Conf. on Genetic Algorithms*, Morgan Kaufmann, San Mateo, CA, 1989, 51.

2

The Walsh Transform and the Theory of the Simple Genetic Algorithm

Michael D. Vose and Alden H. Wright

Abstract   This chapter surveys a number of recent developments concerning the Simple Genetic Algorithm, its formalism, and the application of the Walsh transform to the theory of the Simple Genetic Algorithm.

2.1   Introduction

The Walsh transform has a history of applications to signal processing and pattern recognition. Genetic algorithms are also being successfully applied to problems in these areas. This chapter is about both the Walsh transform and genetic algorithms, though not as they pertain to applications, but as how they relate to each other. Our subject is the mathematical formalization of the Simple Genetic Algorithm and how the Walsh transform appertains to its theoretical development.

We begin with a general description of a heuristic search method, *Random Heuristic Search*, of which the Simple Genetic Algorithm (SGA) is a special case. We then specialize random heuristic search to obtain the SGA. Next the Walsh transform is introduced and we present a series of results which explore its relevance to theoretical aspects of the SGA.

This chapter surveys topics, many of which have appeared in electronic bulletin boards, technical reports, conference presentations, books, or journal articles. They have not, however, been integrated into a coherent account before. Several results will be sketched, by leaving technicalities to the reader, and others will be summarized. Complete details will eventually be available in book form (together with additional material) in *The Simple Genetic Algorithm: Foundations and Theory* to be published by MIT Press next year [4].

2.2   Random Heuristic Search

An instance of random heuristic search can be thought of as an initial collection of elements P_0 chosen from the search space, together with some transition rule τ which from P_i will produce another collection P_{i+1}. In general, τ will be iterated to produce a sequence

$$P_0 \overset{\tau}{\mapsto} P_1 \overset{\tau}{\mapsto} P_2 \overset{\tau}{\mapsto} \dots$$

The beginning collection P_0 is referred to as the *initial population*, the first population (or *generation*) is P_1, the second generation is P_2, and so on. Populations are generated successively until some stopping criteria are reached, when it is hoped that the object of search has been found.

For our purposes, the search space Ω is the set of length ℓ binary strings. Integers in the interval $[0, 2^\ell)$ are identified with elements of Ω through their binary representation. The algorithms which random heuristic search comprise are constrained by which transition rules are allowed. Characterizing admissible τ will be postponed until after the representation scheme for populations has been introduced. Let $n = 2^\ell$ and define the *simplex* to be the set

$$\Lambda \;=\; \{<x_0, ..., x_{n-1}> \,:\, x_j \in \Re,\; x_j \geq 0,\; \Sigma\, x_j = 1\}$$

An element p of Λ corresponds to a population (populations are bags) according to the rule

$$p_i \;=\; \text{the proportion of } i \text{ contained in the population}$$

where indexing always begins at zero. The cardinality of each generation is a constant r called the *population size*. Hence the proportional representation given by p unambiguously determines a population once r is known. The vector p is referred to as a *population vector* (or *descriptor*). For example, the population vector

$$< 0.1,\ 0.2,\ 0.0,\ 0.7 >$$

refers to string length 2, since $n = 4 = 2^\ell$. The population described contains 10% of 0 (string 00), 20% of 1 (string 01), and 70% of 3 (string 11).

Given the current population P, the next population $Q = \tau(P)$ cannot be predicted with certainty because τ is stochastic; Q results from r independent, identically distributed random choices. Let $\mathcal{G} : \Lambda \to \Lambda$ be a function which given the current population vector p produces a vector whose i th component is the probability that i is the result of a random choice. Thus $\mathcal{G}(p)$ is that probability vector which specifies the distribution by which the aggregate of r choices forms the subsequent generation Q. A transition rule τ is admissible if it corresponds to a heuristic function $\mathcal{G}$ in this way.

In terms of search, P is the starting configuration with corresponding descriptor p, and $\mathcal{G}(p)$ is the *heuristic* according to which the search space is to be explored. The result $Q = \tau(P)$ of that exploration is the next generation. A new heuristic $\mathcal{G}(q)$ (here q is the descriptor of Q) may be invoked to repeat the cycle.

Perhaps the first and most natural question concerning random heuristic search is what connection is there between the heuristic function used and the expected next generation? The following theorem provides the answer [3]:

THEOREM 2.1:
Let p be the current population vector, and let $\mathcal{G}$ be the heuristic function used in random heuristic search. The expected next population vector is $\mathcal{G}(p)$.

According to the law of large numbers, if the next generation's population vector q were obtained as the result of an infinite sample from the distribution described by $\mathcal{G}(p)$, then q would match the expectation, hence $q = \mathcal{G}(p)$. Because this corresponds to random heuristic search with an infinite population, the deterministic algorithm resulting from "$\tau = \mathcal{G}$" is called the *infinite population algorithm*.[1]

At this point, random heuristic search and the infinite population algorithm have been defined. The following subsections will specify how the heuristic function $\mathcal{G}$ may be chosen to yield the SGA. The description of $\mathcal{G}$ proceeds through steps analogous to a procedural definition of the simple GA. A subsection is devoted to each (selection, mutation, and crossover). The final subsection summarizes the definition and interpretation of $\mathcal{G}$.

2.2.1 Notation

Square brackets $[\cdots]$ are, besides their standard use as specifying a closed interval of numbers, used to denote an indicator function: if *expr* is an expression which may be true or false, then

$$[expr] \quad = \quad \begin{cases} 1 & \text{if } expr \text{ is true} \\ 0 & \text{otherwise} \end{cases}$$

The delta function is defined by $\delta_{i,j} = [i = j]$.

Since integers in the interval $[0, n)$ are identified with elements of Ω through their binary representation, this correspondence allows them to be regarded as

---

[1]Strictly speaking, τ produces the next generation from the current, while $\mathcal{G}$ produces the *representation* of the expected next generation from the *representation* of the current. This distinction between a population and its representation is conveniently blurred.

elements of the product group

$$\Omega \;\; = \;\; \underbrace{Z_2 \times \cdots \times Z_2}_{\ell \text{ times}}$$

where Z_2 denotes the integers modulo 2. The group operation on the product is denoted by $\oplus$, and the operation of componentwise multiplication is denoted by $\otimes$. Elements of this product will also be thought of as column vectors in $\mathfrak{R}^\ell$.

Indexing of vectors and matrices always begins with zero, and T denotes transpose. Let $\mathbf{1}$ denote the column vector of all 1s. The j th basis vector e_j is the column vector identical to 0 except that it contains 1 in the j th row. The notation $\bar{k}$ abbreviates $k \oplus \mathbf{1}$. The operation $\otimes$ takes precedence over $\oplus$, and both bind more tightly than other operations.

The set of all elements $x \in \Omega$ satisfying $x \otimes \bar{k} = 0$ is denoted by Ω_k. Each $i \in \Omega$ has a unique representation $i = u \oplus v$ where $u \in \Omega_k$ and $v \in \Omega_{\bar{k}}$. This follows by choosing $u = i \otimes k$ and $v = i \otimes \bar{k}$.

2.2.2   Selection

Given a fitness function $f : \Omega \to \mathfrak{R}^+$, define the *fitness matrix* F to be the $n \times n$ diagonal matrix $F_{i,i} = f(i)$. Since f is positive, F is invertible. It follows that the function $\mathcal{F}(x) = Fx/\mathbf{1}^T Fx$ is also invertible and

$$\mathcal{F} : \Lambda \longrightarrow \Lambda$$

The image of a population vector p under $\mathcal{F}$ is called a *selection vector*. The i th component of $\mathcal{F}(p)$ is the probability with which i is to be selected (with replacement) from the current population p.

2.2.3   Mutation

The symbol μ will be used for three different (though related) things. This overloading of μ does not take long to get used to because context makes its meaning clear. The benefits are clean and elegant presentation and the ability to use a common symbol for ideas whose differences are often conveniently blurred.

First, $\mu \in \Lambda$ can be regarded as a distribution describing the probability μ_i with which i is selected to be a mutation mask (additional details follow).

Second, $\mu : \Omega \to \Omega$ can be regarded as a *mutation function* which is nondeterministic. The result $\mu(x)$ of applying μ to x is $x \oplus i$ with probability μ_i. The i occurring in $x \oplus i$ is referred to as a *mutation mask*.

Third, $\mu \in [0, 0.5)$ can be regarded as a *mutation rate* which implicitly specifies the distribution μ according to the rule

$$\mu_i = (\mu)^{\mathbf{1}^T i} (1 - \mu)^{\ell - \mathbf{1}^T i}$$

The distribution μ need not correspond to any mutation rate, although that is certainly the classical situation. Any element $\mu \in \Lambda$ whatsoever is allowed. The effect of mutation is to alter the bits of string x in those positions where the mutation mask i is 1. For arbitrary $\mu \in \Lambda$, mutation is called *positive* if $\mu_i > 0$ for all i. Mutation is called *zero* if $\mu_i = [i = 0]$.

Mutation is called *independent* if for all j and k

$$\mu_j = \sum_{k \otimes i = 0} \mu_{i \oplus j} \sum_{\overline{k} \otimes i = 0} \mu_{i \oplus j}$$

This latter classification is of interest because of the relationship of independent mutation to crossover (to be considered next). When mutation is independent—which is the case when mutation corresponds to a rate—it may be performed either before or after crossover; the effects are exactly the same.

2.2.4 Crossover

It is convenient to use the concept of *partial probability*. Let $\zeta : A \to B$ and suppose $\phi : A \to [0, 1]$. To say "$\xi = \zeta(a)$ with partial probability $\phi(a)$" means, for all b, that ξ takes the value b with probability $\sum [\zeta(a) = b] \phi(a)$.

The description of crossover parallels the treatment given to mutation. The symbol X will be used for three different (though related) things.

First, $X \in \Lambda$ can be regarded as a distribution describing the probability X_i with which i is selected to be a crossover mask (additional details follow).

Second, $X : \Omega \times \Omega \to \Omega$ can be regarded as a *crossover function* which is nondeterministic. The result $X(x, y)$ is $x \otimes i \oplus \overline{i} \otimes y$ with partial probability $X_i/2$ and is $y \otimes i \oplus \overline{i} \otimes x$ with partial probability $X_i/2$. The i occurring in the definition of $X(x, y)$ is referred to as a *crossover mask*.

The arguments x and y of the crossover function are called *parents*, and the pair $x \otimes i \oplus \overline{i} \otimes y$ and $y \otimes i \oplus \overline{i} \otimes x$ are referred to as their *children*. Note that crossover produces children by exchanging the bits of parents in those positions where the crossover mask i is 1.

Third, $X \in [0, 1]$ can be regarded as a *crossover rate* which specifies the distribution X according to the rule

$$X_i = \begin{cases} X c_i & \text{if } i > 0 \\ 1 - X + X c_0 & \text{if } i = 0 \end{cases}$$

where the distribution $c \in \Lambda$ is referred to as the *crossover type*. Classical crossover types include *1-point crossover*, for which

$$c_i = \begin{cases} 1/(\ell - 1) & \text{if } \exists k \in (0, \ell) \text{ and } i = 2^k - 1 \\ 0 & \text{otherwise} \end{cases}$$

and *uniform crossover*, for which $c_i = 2^{-\ell}$. Any element $c \in \Lambda$ whatsoever is allowed as a crossover type.

2.2.5   The Heuristic Function of the Simple Genetic Algorithm

The SGA is given by the heuristic function corresponding to selection (twice) to produce x and y, followed by mutation of x and y, followed by crossover of the results of mutation. The pair selected are called parents, and the end result (only one of the two strings resulting from crossover is kept) is their child. The *mixing matrix M* is defined by the probability that child 0 is obtained:

$$M_{x,y} = \sum_{i,j,k} \mu_i\, \mu_j\, \frac{\chi_k + \chi_{\bar{k}}}{2} [(x \oplus i) \otimes k \oplus \bar{k} \otimes (y \oplus j) = 0]$$

Let σ_k be the permutation matrix with i, j th entry given by $[i \oplus k = j]$ and define the *mixing function* $\mathcal{M} : \Lambda \to \Lambda$ by $\mathcal{M}(x)_i = (\sigma_i x)^T M \sigma_i x$. The i th component function $\mathcal{G}_i$ of the simple genetic algorithm's heuristic function is the probability that i is the end result of selection, mutation, and crossover. In vector form it is

$$\mathcal{G}(p) = \mathcal{M} \circ \mathcal{F}(p)$$

At this stage, the heuristic function which reduces random heuristic search to the SGA has been determined; $\mathcal{G}$ as defined above gives the exact sampling distribution which determines genetic search. To reinforce this point, consider the following algorithm:

1. Generate a random population p.

2. Compute the distribution $\mathcal{G}(p)$.

3. Determine the next generation by sampling strings according to the distribution in step 2.

4. Replace p by the result of step 3 and go to step 2.

THEOREM 2.2:
The previous algorithm is indistinguishable (for all generations, for all string lengths, and for all finite population sizes) from the following SGA:

1. Generate a random population p of fixed length binary strings.

2. Choose two parents u and v from p by proportional selection.

   a. Cross u and v (by any fixed crossover rate and type) to produce children u' and v'.

   b. Mutate u' and v' (by any fixed mutation rate and type) to produce u'' and v''.

   c. Keep (with uniform probability) one of u'' and v'' for the next generation.

3. If the next generation is incomplete, repeat step 2.

4. Replace p by the new generation just formed and go to step 2.

At this point we have seen that the SGA is an instance of random heuristic search and have identified its heuristic function. Moreover, $\mathcal{G}$ provides exact answers to the following fundamental questions:

- What is the *exact sampling distribution* describing the formation of the next generation?

- What is the *expected next generation*?

- In the limit, as population size grows, what is the *transition function* which maps from one generation to the next?

2.3   The Walsh Transform

The *Walsh matrix* is defined by

$$W_{i,j} \;=\; 2^{-\ell/2}(-1)^{i^T j}$$

It is symmetric and orthogonal. The *Walsh transform* is the linear transformation $x \mapsto Wx$. Since $W^{-1} = W$, the Walsh transform is self-inverse.

In order to keep formulas simple, it is helpful, for matrix A and vector v, to represent WAW and Wv concisely. The former is denoted by $\widehat{A}$ and the latter by $\widehat{v}$. The matrix $\widehat{A}$ is referred to as the Walsh transform of the matrix A. Note that the transform of a vector is left multiplication by W, while the transform of a matrix is conjugation by W. This notation is particularly useful since $\widehat{Ax} = \widehat{A}\widehat{x}$. If y is a *row vector*, then $\widehat{y}$ denotes yW. With this convention, we have $\widehat{yA} = \widehat{y}\widehat{A}$, and the Walsh transform commutes with transpose.

Basic properties useful in computing with transforms include

$$i^T j = j^T i$$

$$(-1)^{i^T(j \oplus k)} = (-1)^{i^T(j+k)}$$

$$\sum_i (-1)^{i^T j} = 2^\ell \delta_{j,0}$$

$$\widehat{\sigma}_k = 2^{\ell/2} diag(\widehat{e}_k)$$

The first theorem demonstrates the amazing ability of the Walsh transform to unravel the complexity of mixing. A consequence will be an explicit formula for the spectrum of the differential of the mixing function $\mathcal{M}$. The following theorem is a generalization of a result due to Koehler [1]. His theorem is the special case corresponding to $x = 0$ and one-point crossover.

THEOREM 2.3:

$$\widehat{M}_{x,y} = 2^{\ell-1} \delta_{x \otimes y, 0} \widehat{\mu}_x \widehat{\mu}_y \sum_{k \in \Omega_{\overline{x} \otimes \overline{y}}} \chi_{k \oplus x} + \chi_{k \oplus y}$$

With the representation provided by Theorem 2.3:, it is easy to establish the following.

COROLLARY 2.1:
If mutation is zero, then $WM = MW$.

Define the *twist* A_* of a matrix A by $(A_*)_{i,j} = A_{i \oplus j, i}$. The following result is of key importance.

COROLLARY 2.2:
$\widehat{M}_*$ is lower triangular. If mutation is zero, then M_* is upper triangular.

> Sketch of proof: For any matrix A, the equality $(W A_* W)_{i,j} = (WAW)_{j,i \oplus j}$ follows from expanding both sides. Thus $\widehat{M}_*$ is lower triangular since by Theorem 2.3:, $(\widehat{M})_{j,i \oplus j}$ is nonzero only when $(i \oplus j) \otimes j = 0$, which is equivalent to $\overline{i} \otimes j = 0$, which implies $i \geq j$. When mutation is zero, appealing to Corollary 2.1: gives

$$(\widehat{M_*})_{i,j} = (WMW)_{j,i\oplus j}$$

$$= M_{j,i\oplus j}$$

$$= M_{j\oplus i,j}$$

$$= (M_*)_{j,i}$$

$$= (M_*^T)_{i,j}$$

Since $\widehat{M_*}$ is lower triangular, it follows that M_* is upper triangular.
□

Note what these results have accomplished. Direct access to the spectrum of M_* has been obtained since M_* is triangularized by the Walsh transform. The main application of Theorem 2.3: is to the differential $d\mathcal{M}$ of the mixing function $\mathcal{M}$. Some preliminary observations prepare the way.

Observe that $(\sigma_u M_* \sigma_u)_{i,j} = (M_*)_{i\oplus u, j\oplus u} = M_{i\oplus j, u\oplus i}$. Next note that the i, j th entry of $d\mathcal{M}_x$ is given by

$$\frac{\partial}{\partial x_j} \sum_{u,v} x_u x_v M_{u\oplus i, v\oplus i} = \sum_{u,v} (\delta_{u,j} x_v + \delta_{v,j} x_u) M_{u\oplus i, v\oplus i}$$

$$= 2 \sum_u x_u M_{i\oplus j, u\oplus i}$$

This establishes the following:

THEOREM 2.4:
$$d\mathcal{M}_x = 2 \sum \sigma_u M_* \sigma_u x_u$$

PROPOSITION 2.1:
$$(\widehat{d\mathcal{M}_x})_{i,j} = 2^{1+\ell/2} (\widehat{M_*})_{i,j} \widehat{x}_{i\oplus j}$$

Sketch of proof: Using the formula $(\widehat{\sigma_k})_{i,j} = (-1)^{j^T k} \delta_{i,j}$,

$$(\widehat{d\mathcal{M}_x})_{i,j} = 2 \sum_k (\widehat{\sigma_k} \widehat{M_*} \widehat{\sigma_k})_{i,j} x_k$$

$$= 2 \sum_k x_k \sum_{u,v} (-1)^{u^T k} \delta_{i,u} (\widehat{M_*})_{u,v} (-1)^{j^T k} \delta_{v,j}$$

$$= 2 \sum_k x_k (-1)^{k^T(i+j)} (\widehat{M_*})_{i,j}$$

$$= 2^{1+\ell/2} (\widehat{M_*})_{i,j} (Wx)_{i \oplus j} \qquad\qquad \square$$

The following result from linear algebra will be useful:

PROPOSITION 2.2:
Suppose $Ax = \lambda x$. Then

$$\mathrm{spec}(A) \quad = \quad \mathrm{spec}(A^T \Big|_{x^\perp}) \cup \{\lambda\}$$

We now sketch the main result of this section.

THEOREM 2.5:
The spectrum of $d\mathcal{M}_x$ is the spectrum of M_* multiplied by $2 \cdot 1^T x$. In particular, for $x \in \Lambda$, it is independent of x and is given by 2 times the 0th column of $\widehat{M}$. If, moreover, mutation is positive, then the largest eigenvalue is 2 and all other eigenvalues are in the interior of the unit disk.

Sketch of proof: Because the spectrum is invariant under conjugation by W (a change of basis), it suffices to consider the spectrum of $\widehat{d\mathcal{M}_x}$. By Corollary 2.2: and Proposition 2.1:, $\widehat{d\mathcal{M}_x}$ is lower triangular. Hence the spectrum is given by the diagonal entries. Note that $2^{1+\ell/2} (\widehat{M_*})_{i,i} \, \widehat{x}_0 = 2 \widehat{M}_{i,0} 1^T x$.
Since $(M_*)^T 1 = 1$, it follows from Proposition 2.2: that

$$\mathrm{spec}(M_* \Big|_{1^\perp}) \quad = \quad \mathrm{spec}(M_*) \setminus \{1\}$$

Combining this with Theorem 2.3: shows the spectral radius of M_* restricted to $1^\perp$ is

$$\frac{1}{2} \sup_{j>0} (2^{\ell/2} W\mu)_j \sum_{k \in \Omega_{\bar{j}}} \chi_k + \chi_{k \oplus j}$$

If μ is positive and $j > 0$, cancellation occurs in the sum defining $(2^{\ell/2} W \mu)_j$ and so it must have absolute value less than 1. Next note that the subscripts in the sum above are of the form u and $v \oplus j$ where $u, v \in \Omega_{\bar{j}}$. Since $\Omega_{\bar{j}}$ is a group, $u = v \oplus j$ is impossible; it would lead to the contradiction $u \oplus v = j \in \Omega_{\bar{j}}$. The sum can therefore have no repeated terms and is at most 1. Hence the spectral radius of M_* restricted to $\mathbf{1}^{\perp}$ is less than $\frac{1}{2}$. $\square$

To summarize main results:

- Theorem 2.3: shows how the Walsh transform simplifies M.

- Corollary 2.2: shows the Walsh transform triangulates M_*.

- A consequence is Theorem 2.5: which reveals the spectrum of $d\mathcal{M}$ as a column of $2\widehat{M}$.

2.4 The Walsh Basis

The hyperplane containing Λ is a translate in the direction of the first column of W of the linear span of the other columns. This observation suggests a natural basis for representing how $\mathcal{G}$ transforms space is $\mathcal{B} = \{\widehat{e_0}, ..., \widehat{e_{n-1}}\}$. The development of how $\mathcal{G}$ transforms in this representation is the subject of this section.

THEOREM 2.6:

$$d\mathcal{M}_{\widehat{e_j}} \widehat{e_i} = 2^{1+\ell/2} \widehat{M}_{i,j} \widehat{e_{i \oplus j}}$$

and

$$\widehat{e_i}^T d\mathcal{M}_{\widehat{e_j}} = 2^{1+\ell/2} \widehat{M}_{i \oplus j, j} \widehat{e_{i \oplus j}}^T$$

Sketch of proof: By Proposition 2.1:,

$$d\mathcal{M}_{\widehat{e_j}} \widehat{e_i} = W(W d\mathcal{M}_{\widehat{e_j}} W) e_i$$

$$= W(\widehat{d\mathcal{M}_{\widehat{e_j}}} e_i)$$

$$= 2^{1+\ell/2} W \sum_k (\widehat{M_*})_{k,i} \, \delta_{k \oplus i, j} \, e_k$$

$$= 2^{1+\ell/2} \, (\widehat{M_*})_{i \oplus j, i} \; W e_{i \oplus j}$$

The second equation is a consequence of the first.   □

THEOREM 2.7:
$d\mathcal{M}_x \, y$ is symmetric and linear in x and y. Moreover,

$$\mathcal{M}(x) = \frac{1}{2} d\mathcal{M}_x \, x$$

$$\mathcal{M}(x) - \mathcal{M}(y) = d\mathcal{M}_{\frac{x+y}{2}} \, (x - y)$$

$$\mathbf{1}^T d\mathcal{M}_x = 2 \, \mathbf{1}^T x \, \mathbf{1}^T$$

$$\mathbf{1}^T \mathcal{M}(x) = (\mathbf{1}^T x)^2$$

Sketch of proof: Symmetry in x and y follows from linearity and the fact that symmetry holds on a basis (Theorem 2.6:). Linearity is a consequence of Theorem 2.4:. The first formula follows from expanding both sides, and the second is a consequence of symmetry, linearity, and the first. The third formula follows from linearity, the second formula of Theorem 2.6:, and Theorem 2.3:. The last formula follows from the third and first.   □

We are now positioned to derive how $\mathcal{G}$ transforms in the coordinates corresponding to $\mathcal{B}$. Multiplying the representation

$$x = \sum x_j e_j$$

by W and then replacing x by $\widehat{x}$ in the resulting expression yields

$$x = \sum \widehat{x_j} \widehat{e_j}$$

The ability to pass between these two representations for x will be useful. By Theorem 2.7: we can write $\mathcal{M}(x)$ as

$$\frac{1}{2} d\mathcal{M}_x x$$

Using the second representation for x given above and expanding by the bilinearity of $dM_{(\cdot)}(\cdot)$ allows this to be written as

$$\frac{1}{2} \sum_{i,j} \widehat{x}_i \widehat{x}_j dM_{\widehat{e_i}} \widehat{e_j}$$

Appealing to the first formula of Theorem 2.6: and making the change of variables $i \oplus j = k$ leads to

$$M(x) = 2^{\ell/2} \sum_k \widehat{e_k} \sum_i \widehat{x}_i \widehat{x}_{i \oplus k} \widehat{M}_{i,i \oplus k}$$

This derivation together with the observation that $\widehat{M}_{i,i \oplus k} \neq 0 \implies i \in \Omega_k$ (a consequence of the δ factor given in Theorem 2.3:) establishes the following:

THEOREM 2.8:
The k th component of $M(x)$ with respect to the basis B is

$$2^{\ell/2} \sum_{i \in \Omega_k} \widehat{x}_i \widehat{x}_{i \oplus k} \widehat{M}_{i,i \oplus k}$$

The next step is to calculate how $\mathcal{F}$ transforms in the coordinates corresponding to B. Observe that

$$Fx = \sum x_j F e_j$$

$$= \sum (Fx)_j e_j$$

$$= \sum (\widehat{Fx})_j \widehat{e_j}$$

Therefore

$$1^T Fx = \sum (\widehat{Fx})_j 1^T \widehat{e_j}$$

$$= 2^{\ell/2} \sum (\widehat{Fx})_j \widehat{e_0}^T \widehat{e_j}$$

$$= 2^{\ell/2} (\widehat{Fx})_0$$

This leads to the next theorem.

THEOREM 2.9:
Let $F = diag(f)$ be the fitness matrix. The k th component of $\mathcal{F}(x)$ with respect to the basis $\mathcal{B}$ is

$$2^{-\ell/2} \widehat{f}^T \sigma_k \widehat{x} / \widehat{f}^T \widehat{x}$$

Sketch of proof: Since $(\widehat{Fx})_j = e_j^T \widehat{Fx}$, the discussion preceding Theorem 2.9: shows it suffices that

$$e_j^T \widehat{F} = 2^{-\ell/2} \widehat{f}^T \sigma_j$$

This follows from

$$e_j^T \widehat{diag(f)} \, WW = \widehat{e_j}^T diag(f) \, W$$

$$= f^T diag(\widehat{e_j}) W$$

$$= \widehat{f}^T \widehat{diag(\widehat{e_j})}$$

$$= \widehat{f}^T 2^{-\ell/2} \sigma_j$$

$\square$

Theorems 2.8: and 2.9: show that computing $\mathcal{G}$ in Walsh coordinates (i.e., computing with respect to the basis $\mathcal{B}$) is far more efficient than computing it in the standard basis. For example, the i th component of $\mathcal{M}(x)$ involves the quadratic form

$$(\sigma_i x)^T M \sigma_i x$$

With positive mutation, M is dense and each computation is $O(n^2)$. Moreover, there are n such components to consider. In comparison, the cost of the corresponding component in Walsh coordinates is the size of Ω_k (by Theorem 2.8:). Hence the total expense is bounded by the order of

$$\sum_k 2^{1^T k} = \sum_u 2^u \sum_k [1^T k = u]$$

$$= \sum_u 2^u \binom{\ell}{u}$$

$$= n^{\log_2 3}$$

This compares favorably with $O(n^3)$. In Walsh coordinates, selection is the dominant cost. By Theorem 2.9: it is $O(n^2)$. Selection in standard coordinates is $O(n)$.

There are reasons far more profound than efficiency, $O(n^2)$ vs. $O(n^3)$, that the Walsh basis $\mathcal{B}$ appertains to the simple genetic algorithm. The next section explains how Walsh coordinates induce a decomposition of Λ into regions of space which are invariant under the mixing function.

2.5 Invariance

We mean by "mixing" the combined operations of crossover and mutation. In particular, when the crossover rate is zero, mixing corresponds to mutation alone. The Walsh basis appertains to the simple genetic algorithm in that it is particularly revealing of the dynamics of mixing.

If p is a population vector which does not have components in the direction of every basis vector, then p represents a population in which some string types are missing. In particular, string i is missing from the population exactly when $p_i = 0$. After mixing, however, *every* string is expected to be represented, provided that mutation is positive, since any string has a nonzero probability of being produced by mutation and surviving crossover (being crossed with itself).

Hence there is no proper subset of the basis vectors whose linear span is invariant under $\mathcal{M}$. If some components of p were zero, those components become positive in the vector $\mathcal{M}(p)$. This is perhaps intuitive, since mutation "spreads out" the initial population to contain—in expectation—instances of every string type.

Nevertheless, in Walsh coordinates there are invariant subspaces—exponentially many of them—even when mutation is positive. If S is a set of vectors, let $\mathcal{L}S$ denote their linear span. The invariance theorem is the following:

THEOREM 2.10:
For all k, both $\mathcal{L}\{\widehat{e_i} : i \in \Omega_k\}$ and $\mathcal{L}\{\widehat{e_j} : j \notin \Omega_k\}$ are invariant under $\mathcal{M}$.

> Sketch of proof: We consider the second space first. Let $x = \sum \alpha_i \widehat{e_i}$ be an element of $\mathcal{L}\{\widehat{e_j} : j \notin \Omega_k\}$. Thus $i \in \Omega_k \Longrightarrow \alpha_i = 0$. By Corollary 2.7: and Theorem 2.6:, for such i

$$\widehat{e_i}^T \mathcal{M}(x) = \frac{1}{2}\widehat{e_i}^T d\mathcal{M}_x x$$

$$= 2^{\ell/2} \sum_{u,v} \alpha_u \alpha_v \widehat{M}_{u,v} \widehat{e_i}^T \widehat{e_{u \oplus v}}$$

$$= 2^{\ell/2} \sum_u \alpha_u \, \alpha_{u \oplus i} \, \widehat{M}_{u,u \oplus i}$$

It follows from Theorem 2.3: that $\widehat{M}_{u,u \oplus i} = 0$ unless $u \in \Omega_i$. Since $\Omega_i \subset \Omega_k$, the coefficients α_u are zero.

Next let $x = \sum \alpha_i \widehat{e_i}$ be an element of $\mathcal{L} \{\widehat{e_i} : i \in \Omega_k\}$. For $j \notin \Omega_k$ we have as before,

$$\widehat{e_j}^T \mathcal{M}(x) \;\; = \;\; 2^{\ell/2} \sum_u \alpha_u \, \alpha_{u \oplus j} \, \widehat{M}_{u,u \oplus j}$$

where nonzero terms are subscripted by elments of Ω_j. Since $j \notin \Omega_k \implies \alpha_j = 0$, every term will be zero provided that

$$u \in \Omega_j \quad \implies \quad u \notin \Omega_k \; \vee \; u \oplus j \notin \Omega_k$$

This implication follows from the fact that $u \oplus (u \oplus j) = j \notin \Omega_k$.
$\square$

Space has, for every choice of $k \in \Omega$, the orthogonal decomposition

$$\mathcal{L} \{\widehat{e_i} : i \in \Omega_k\} \times \mathcal{L} \left\{\widehat{e_j} : j \notin \Omega_k\right\}$$

Theorem 2.10: shows each factor space is invariant under mixing. Since mixing preserves Λ, the intersection of these spaces with Λ is also invariant. As a special case, each region is invariant under mutation.

The representation provided by Theorem 2.8: was essentially responsible for the invariance result of this section; it played a crucial role in the proof of Theorem 2.10:. That same representation can be used to determine the inverse of $\mathcal{M}$. This is done in the next section.

2.6   The Inverse GA

For the remainder of the chapter, we will consider operators $\mathcal{M}$, $\mathcal{F}$, and $\mathcal{G}$ with respect to the basis $\mathcal{B}$. Therefore, a vector x will have i th component $x_i = x^T \widehat{e_i}$ (which previously was denoted $\widehat{x_i}$ in the sections above where the standard basis was used to determine coordinates).

Theorem 2.8: effectively triangulates the equations which relate one generation to the next. Let x be the descriptor of the current generation and let Ω'_k denote

$\Omega_k \setminus \{0, k\}$. The vector $y = \mathcal{M}(x)$ satisfies

$$
y_k \;=\; \begin{cases} x_k & \text{if } k = 0 \\[2mm] 2\widehat{M}_{0,k} x_k \;+\; 2^{\ell/2} \sum_{i \in \Omega'_k} x_i x_{i \oplus k} \widehat{M}_{i,i \oplus k} & \text{if } k > 0 \end{cases}
$$

This relationship follows directly from Theorem 2.8: and the observation that $x_0 = 2^{-\ell/2}$ for all $x \in \Lambda$. Solving for x_k gives the equations

$$
x_k \;=\; \begin{cases} y_k & \text{if } k = 0 \\[2mm] (2\widehat{M}_{0,k})^{-1}(y_k \;-\; 2^{\ell/2} \sum_{i \in \Omega'_k} x_i x_{i \oplus k} \widehat{M}_{i,i \oplus k}) & \text{if } k > 0 \end{cases}
$$

These equations serve to define x_k recursively in terms of x_i for $i \in \Omega'_k$. The recursion terminates in $x_0 = y_0$. Therefore the map

$$
y \mapsto x
$$

determined by this recursive definition is the inverse function to $\mathcal{M}$ on Λ. The inverse is defined provided division by zero is avoided, i.e., $\widehat{M}_{0,k} \neq 0$. By Theorem 2.5:, this is equivalent to the invertibility of M_*.

It is easily verified that the inverse to $\mathcal{F}$ on Λ is

$$
x \;\mapsto\; \widehat{F^{-1}x} \,/\, \widehat{\mathbf{1}^T F^{-1} x}
$$

and so the inverse of $\mathcal{G}$ is given by $\mathcal{F}^{-1} \circ \mathcal{M}^{-1}$.

Having access to the inverse of $\mathcal{G}$ enables the GA to be run in reverse, from one generation to the *previous*. What this means is that $\mathcal{G}^{(-1)}(x)$ is that population y with the property that if the initial population were y then the expected next generation would be x. This holds for all population sizes, finite and infinite.

Since Λ is invariant under $\mathcal{G}$, the only way to leave it—assuming one begins there and is interested in tracing the trajectory of expected evolution—is by moving *backwards* in time through previous generations. Running the discrete dynamical system $\mathcal{E}$ governed by the transition function $\mathcal{G}$ backwards from Λ seems to converge to a point α located outside Λ. Visualization of the dynamics of $\mathcal{E}$ in the neighborhood of α leads to fractal structures. The reader is referred to Reference [2] for further information on this subject.

2.7   Recombination Limits

This section considers the effect of iterated mixing on an arbitrary initial population x. The triangular form for the recombination equations provided by Theorem 2.8: allows the fixed points of $\mathcal{M}$ to be determined *explicitly*. In fact, the triangular form can be used to show that

$$x \in \Lambda \implies \lim_{j \to \infty} \mathcal{M}^{(j)}(x)$$

exists. An important preliminary result (the proof is straightforward) is the following:

LEMMA 2.1:
If $z > 0$ then $|\widehat{M}_{0,z}| \le 1/2$. When equality holds and $z = x \oplus y$ for nonzero x and y, then $\widehat{M}_{x,y} = 0$.

Recall that $y = \mathcal{M}(x)$ satisfies

$$y_k = \begin{cases} x_k & \text{if } k = 0 \\ 2\widehat{M}_{0,k}x_k + \sqrt{n}\sum_{i \in \Omega'_k} x_i x_{i \oplus k} \widehat{M}_{i,i \oplus k} & \text{if } k > 0 \end{cases}$$

where Ω'_k denotes $\Omega'_k \setminus \{0, k\}$. This can be put in the abbreviated form

$$y_k = \begin{cases} x_k & \text{if } k = 0 \\ \alpha_k x_k + \beta_k & \text{if } k > 0 \end{cases}$$

It is easy to see that $\lim_j \mathcal{M}^{(j)}(x)$ exists by considering k th components in order of increasing $1^T k$. But first note when $1^T k = 1$ the sum β_k is empty, hence zero. Also, Lemma 2.1: implies that $|\alpha_k| \le 1$ and that $|\alpha_k| = 1 \implies \beta_k = 0$ provided $1^T k > 1$ (because of the factors $\widehat{M}_{i,i \oplus k}$ in the terms of β_k).

Clearly the 0 th component of $\lim_j \mathcal{M}^{(j)}(x)$ has already converged since $y_0 = x_0$. Next, for $1^T k = 1$, if $\alpha_k = 1$ then $y_k = x_k$ so the k th component has converged. If $\alpha_k = -1$ then the k th component is periodic, oscillating between $\pm x_k$. When $|\alpha_k| < 1$, the k th component converges to 0 because the relation between successive iterations is $y_k = \alpha_k x_k$.

In the general case of $1^T k > 1$, the sum β_k may be treated as having already converged because it involves components subscripted by j with $1^T j < 1^T k$. If for some values of j components are periodic, replacing $\mathcal{M}$ by $\mathcal{M}^2$ restores convergence since oscillations have period 2. If $\alpha_k = 1$, then $y_k = x_k$ (since $\beta_k = 0$) so the k th component has already converged. If $\alpha_k = -1$ then $y_k = -x_k$ and the k th component oscillates for $\mathcal{M}$, but has already converged for $\mathcal{M}^2$. When $|\alpha_k| < 1$, the k th component converges to $\beta_k/(1 - \alpha_k)$ since the relation between successive iterations is $y_k = \alpha_k x_k + \beta_k$.

This discussion is summarized by the following theorem.

THEOREM 2.11:

The function $x \mapsto y$ defined recursively by

$$
y_k = \begin{cases} x_k & \text{if } k = 0 \text{ or } |\widehat{M}_{0,k}| = 1/2 \\ \sqrt{n}(1 - 2\widehat{M}_{0,k})^{-1} \sum_{i \in \Omega_k'} y_i y_{i \oplus k} \widehat{M}_{i,i \oplus k} \end{cases}
$$

otherwise produces the fixed point $y = \lim_j \mathcal{M}^{2j}(x)$ of $\mathcal{M}^2$. If no $\widehat{M}_{0,k}$ equals $-1/2$, then y is also the fixed point $\lim_j \mathcal{M}^j(x)$ of $\mathcal{M}$. Otherwise, $\mathcal{M}^j(x)$ converges to a periodic orbit which oscillates between y and $\mathcal{M}(y)$.

Combining Theorem 2.11: with Theorem 2.5:, we see that the fixed points of $\mathcal{M}$ form a manifold of dimension equal to the multiplicity that $1/2$ has as an eigenvalue of M_*. In particular, if the spectrum of M_* does not contain $1/2$, then $\mathcal{M}$ has the unique fixed point $2^{-\ell/2}\mathbf{1}$.

2.8 Conclusion

This chapter brings together a number of theoretical results demonstrating connections between the Walsh transform and the SGA.

The major connections are through:

- Simplification of the mixing matrix M

- The spectrum of the twist M_* and of the differential $d\mathcal{M}$ of mixing

- How selection $\mathcal{F}$ and mixing $\mathcal{M}$ transform in the Walsh basis $\mathcal{B}$

- Orthogonal decompositions of space invariant under mixing

- The inverse operators $\mathcal{M}^{-1}$ and $\mathcal{G}^{-1}$

- Recombination limits

Acknowledgments

This research was supported by the National Science Foundation: IRI-8917545, IRI-9224917, and CDA-9115428.

References

[1] Koehler, G., A proof of the Vose-Liepins conjecture, *Ann. Math. Artif. Intelligence,* 10, 1994, 409–422.

[2] Juliany, J. and Vose, M.D., The genetic algorithm fractal, in *Evolutionary Computation,* 2(2), 1994, 165–180.

[3] Vose, M.D. and Wright, A.H., Simple genetic algorithms with linear fitness, *Evolutionary Computation,* 2(4), 1995, 347–368.

[4] *The Simple Genetic Algorithm: Foundations and Theory,* MIT Press, Cambridge, in press.

3

Adaptation in Genetic Algorithms

Lalit M. Patnaik and Srinivas Mandavilli

Abstract   Genetic algorithms (GAs) have emerged as effective search and optimization methods with applications in several problem domains. When the underlying search space has several locally optimal solutions apart from the globally optimal solution (i.e., the search space is multimodal), GAs emerge as worthy alternatives to traditional optimization techniques. For several years since their inception in 1975, GAs have been molded in the form proposed by Holland, characterized by constant control parameters and fixed length encodings. Recent research has led to variations in the basic GA mechanism. New selection, mutation, and crossover strategies, distributed and parallel implementations, and adaptive mechanisms to modify the control parameters have been proposed. The control parameters of a GA-crossover probability, mutation probability, and population size-critically control the performance of GAs. In the last few years, several researchers have experimented with adaptive mechanisms to dynamically vary the control parameters to improve the performance of GAs. The success that they have achieved in their pursuits makes it worthwhile to survey strategies for adapting the control parameters. In this chapter, we briefly review recent work on adaptive strategies for modifying control parameters of GAs. Next we discuss in detail our own efforts in this direction which have led to the genesis of the *Adaptive Genetic Algorithm*, a very effective GA variant for multimodal optimization.

3.1   Introduction

Genetic Algorithms [1, 6, 7, 12, 20] (GAs) are robust search and optimization techniques which are finding application in a number of practical problems. The robustness of GAs is due to their capacity to locate the global optimum in a multimodal landscape. A plethora of such multimodal functions exist in engineering problems: optimization of neural network structure and learning neural network weights, solving optimal control problems, designing structures, and solving flow problems are a few examples. It is for the above reason that considerable attention has been paid to the design of GAs for optimizing multimodal functions.

GAs employ a random, yet directed search for locating the globally optimal solution. They are superior to "gradient descent" techniques as the search is not biased toward the locally optimal solution. On the other hand, they differ from random sampling algorithms due to their ability to direct the search toward relatively "prospective" regions in the search space.

Typically a GA is characterized by the following components:

- A genetic representation (or an encoding) for the feasible solutions to the optimization problem

- A population of encoded solutions

- A fitness function that evaluates the optimality of each solution

- Genetic operators that generate a new population from the existing population

- Control parameters

The GA may be viewed as an evolutionary process wherein a population of solutions evolves over a sequence of generations. During each generation, the fitness of each solution is evaluated, and solutions are *selected* for reproduction based on their fitness. *Selection* embodies the principle of "survival of the fittest". "Good" solutions are *selected* for reproduction while "bad" solutions are eliminated. The "goodness" of a solution is determined from its fitness value. The selected solutions then undergo recombination under the action of the *crossover* and *mutation* operators. It has to be noted that the genetic representation may differ considerably from the natural form of the parameters of the solutions. Fixed-length and binary-encoded strings for representing solutions have dominated GA research.

The power of GAs arises from *crossover*. Crossover causes a structured, yet randomized exchange of genetic material between solutions, with the possibility that "good" solutions can generate "better" ones.

Crossover occurs only with some probability p_c (the crossover probability or crossover rate). When the solutions are not subjected to crossover, they remain unmodified. Notable crossover techniques include the single-point, the two-point, and the uniform types [21].

Mutation involves the modification of the value of each "gene" of a solution with some probability p_m (the mutation probability). The traditional role of mutation in GAs has been that of restoring lost or unexplored genetic material into the population to prevent the premature convergence of the GA to suboptimal solutions.

However, recent investigations have demonstrated that high levels of mutation could form an effective search strategy when combined with conservative selection methods.

Apart from selection, crossover, and mutation, various other auxiliary operations are common in GAs. Of these, *scaling* mechanisms [11] are widely used. *Scaling* involves a readjustment of fitness values of solutions to sustain a steady selective pressure in the population, and to prevent the premature convergence of the population to suboptimal solutions.

The basic structure of a GA is illustrated in Figure 3.1.

```
Simple Genetic Algorithm ()
{
initialize population;
evaluate population ;
while convergence not achieved
        {
        scale population fitnesses ;
        select solutions for next population ;
        perform crossover and mutation ;
        evaluate population ;
        }
}
```

FIGURE 3.1
Basic structure of a GA.

The chapter is organized as follows: in Section 3.2 we discuss the tradeoffs between exploration and exploitation that a GA needs to make to optimize its search; we analyze how adaptive strategies for dynamically modifying the control parameters can lead to improved GA-search. In Section 3.3 we discuss the motivating factors for employing adaptive control parameters. Section 3.4 describes our approach of using adaptively varying probabilities of crossover and mutation for multimodal function optimization. In Section 3.5, we present experimental results to compare the performance of the GAs with and without adaptive probabilities of crossover and mutation. The conclusions and directions for future work are presented in Section 3.6.

3.2 Exploitation vs. Exploration in Genetic Algorithms

The functioning of GAs may be visualized as a balanced combination of exploration of new regions in the search space and exploitation of already sampled regions. The balance between exploitation and exploration, which critically controls the performance of GAs, is determined by the right choice of control parameters—the crossover and mutation rates and the population size.

The choice of the optimal control parameters for GAs has been a highly debated issue, with both analytical and empirical investigations devoted to it. Some of the tradeoffs that arise with the choice of the optimal control parameters are determined by the following issues:

- Increasing the crossover probability increases recombination of building blocks while it also increases the disruption of good strings.

- Increasing the mutation probability tends to transform the genetic search into a random search while it also aids in reintroducing lost genetic material into the population.

- Increasing the population size increases the diversity in the population and the probability that the GA does not prematurely converge to a local optimum, while it also increases the time required for the population to converge to the optimal regions in the search space.

More importantly, the control parameters have to be chosen after considering the interactions between the genetic operators. They cannot be determined independently, and the choice of the control parameters itself has been viewed as a complex nonlinear optimization problem. Further, it is becoming evident that the optimal control parameters critically depend on the nature of the objective function.

Thus, the choice of the optimal control parameters for GAs largely remains an open issue to date. However, several researchers have proposed control parameter sets that guarantee good performance on carefully chosen testbeds of objective functions. Two distinct parameter sets have emerged to be common in GA practice. One has a small population size and relatively large mutation and crossover probabilities, while the other has a larger population size, but much smaller crossover and mutation probabilities. Typical of these two categories are [crossover rate (p_c): 0.6, mutation rate (p_m): 0.001, population size (P): 100] (due to DeJong) and [$p_c = 0.9$, $p_m = 0.01$, $P = 30$] (due to Grefenstette). While the first set of parameters clearly highlights the secondary role of mutations in GA search, the second set of parameters points to a more significant role for mutations. The high crossover rate of 0.9 suggested in the second set also indicates that a high level of disruption of strings is desirable in small populations.

3.3 Why Adapt Control Parameters?

In the previous section, we have highlighted the critical role of control parameters in controlling GA performance. However, it is not easy to choose optimal control parameters for a given problem. Even to this date, there exist no reliable guidelines to choose control parameters for a given problem. Considerable effort needs to be spent on determining the optimal control parameters, as it involves extensive experimentation. Adaptive strategies are attractive as they partially free the user from this arduous task. The adaptive strategy could quickly adjust the initially defined control parameters to make them optimal for maximizing the performance of the GA.

The second problem with static control parameters is more profound. A control parameter set that is optimal during the initial stages of a search is typically not effective in the later stages. For example, consider the GA search being performed on a multimodal landscape. Initially, we would like the GA to locate the region of the globally optimal solution. High values of crossover probability and mutation probability are suitable since they aid the GA in escaping from local optima. However, in the later stages, when the region of the global optimum is located, the GA should "creep" toward the optimal solution. The crossover and mutation probabilities should ideally be small during this phase. With adaptive strategies for varying the control parameters, it may be possible to readjust the control parameters temporally to maximize performance.

These advantages have motivated researchers to consider a variety of adaptive strategies to dynamically vary control parameters in GAs. Associated with an adaptive mechanism for varying a control parameter is an input metric of the state of the population that triggers the adaptation. The metric could be an explicit estimate of the transient performance of the GA, or an implicit abstract measure that controls performance. Let us consider two examples. An explicit metric could be the increase in the fitness of the best individual in a fixed number of generations. An implicit metric could be the variance of fitness in the population.

The strategy used for adapting the control parameters depends on the definition of the performance of the GA. Again we use examples for illustration. In a nonstationary environment, where the optimal solution changes with time, the GA should possess the capacity to track the optimal solution, too. The adaptation strategy needs to vary the control parameters appropriately whenever the GA is not able to track the located optimum. In another situation, the performance of the GA might be characterized by the maximum fitness of solutions in the population. There, the GA needs to consistently maintain a good solution in the population. Given the performance criterion, an adaptive strategy needs to be designed to realize the specific goal.

With this brief overview, we now review from the literature some strategies for adapting control parameters. Later we discuss in detail our implementation of a

GA that adapts the crossover and mutation probabilities based on the fitnesses in the population.

Davis [2, 3] discusses an effective method of adapting operator probabilities based on the performance of the operators. The adaptation strategy is discussed in the context of a steady-state GA wherein exactly one offspring is produced every iteration. If the offspring is better than the worst solution in the population, it replaces the worst solution. Thus the best solutions that have been located by the GA are "saved" in the population. One of five different recombination operators are used to generate the offspring—two crossover operators and three mutation operators. The purpose of adaptation is to arrive at the right distribution of operator probabilities that maximizes the capacity of the GA to locate better solutions.

The adaptation mechanism provides for the alteration of operator probabilities in proportion to the fitnesses of strings created by the operators. Simply stated, operators which create and cause the generation of better strings are allotted higher probabilities. Also, operators that generate the ancestors of good solutions are rewarded. An explicit metric—fitness—is employed to evaluate the operators. Based on the relative performance of the operators, the probabilities of their usage are adapted periodically. Experiments on two simple objective functions reveal that the distribution of operator probabilities continuously changes over the execution period of the GA. The authors have also demonstrated that the proposed adaptive approach delivers better results than a GA with optimized static operator probabilities. An added advantage is the elimination of the compute-intensive stage for arriving at the static operator probabilities.

In an approach employing a form of adaptive mutation (GENITOR), Whitley and Starkweather [22] have reported significant performance improvements. The probability of mutation is a dynamically varying parameter determined from the Hamming distance between the parent solutions. The diversity in the population is sustained by subjecting similar solutions to increased levels of mutation. Here, too, a steady-state GA is employed with the best solutions retained in the population. In this case, the authors have used an implicit metric—the diversity between two solutions to be crossed—to drive the adaptation. While the goal is to maintain the diversity in the population, the idea behind this adaptation strategy is that the GA's performance is sustained by generating diversity in the population.

Cobb [13] has investigated an adaptive hypermutation operator to enhance the GA's efficacy in nonstationary environments. The adaptation consists of altering the mutation rate between two extreme values of 0.001 and 0.5. The mutation rate is steeply increased during periods of nonstationarity, while it is kept at a low level of 0.001 in stationary periods. Experiments include changing the rate of variation of the objective function characteristics and evaluating the performance of the GA with a wide range of static mutation probabilities. The time-averaged performance of the hypermutation operator has been shown to be significantly better than that of the best static mutation probability.

Smith [17] has proposed an adaptive method for resizing the population based on controlling the fitness-variance of schemata. The metric employed is the variance of schemata in the population. The adaptation policy is to resize the population such that samples of the schemata in the population provide reasonable estimates of the actual variance in the schema fitness.

In the rest of the chapter, we discuss in detail an adaptive strategy for varying the crossover and mutation probabilities that we have proposed [19]. The adaptation is based on the fitness values of strings, and it involves modifying the crossover and mutation probabilities to improve search in multimodal landscapes.

3.4 Adaptive Probabilities of Crossover and Mutation

3.4.1 Motivations

It is essential to have two characteristics in GAs for optimizing multimodal functions. The first characteristic is the capacity to converge to an optimum (local or global) after locating the region containing the optimum. The second characteristic is the capacity to explore new regions of the solution space in search of the global optimum. The balance between these characteristics of the GA is dictated by the values of p_m and p_c and the type of crossover employed [21]. Increasing values of p_m and p_c promote exploration at the expense of exploitation. Moderately large values of p_c (0.5 to 1.0) and small values of p_m (0.001 to 0.05) are commonly employed in GA practice. In our approach, we aim at achieving this tradeoff between exploration and exploitation in a different manner, by varying p_c and p_m adaptively in response to the fitness values of the solutions; p_c and p_m are increased when the population tends to get stuck at a local optimum, and are decreased when the population is scattered in the solution space.

3.4.2 Design of Adaptive p_c and p_m

In order to vary p_c and p_m adaptively, for preventing premature convergence of the GA to a local optimum, it is essential to be able to identify whether the GA is converging to an optimum. One possible way of detecting convergence is to observe the average fitness value $\bar{f}$ of the population in relation to the maximum fitness value f_{max} of the population. $f_{max} - \bar{f}$ is likely to be less for a population that has converged to an optimum solution than that for a population scattered in the solution space. We use the difference in the average and maximum fitness values, $f_{max} - \bar{f}$, as a yardstick for detecting the convergence of the GA. The values of p_c and p_m are varied depending on the value of $f_{max} - \bar{f}$. Since p_c and p_m have to be increased when the GA converges to a local optimum, i.e., when $f_{max} - \bar{f}$ decreases, p_c and p_m will have to be varied inversely with $f_{max} - \bar{f}$.

The expressions that we have chosen for p_c and p_m are of the form

$$p_c = k_1/(f_{max} - \overline{f})$$

and

$$p_m = k_2/(f_{max} - \overline{f}).$$

It has to be observed that, in the above expressions, p_c and p_m do not depend on the fitness value of any particular solution and have the same values for all the solutions of the population. Consequently, solutions with high fitness values as well as solutions with low fitness values are subjected to the same levels of mutation and crossover. When a population converges to a globally optimal solution (or even a locally optimal solution), p_c and p_m increase and may cause the disruption of the near-optimal solutions. The population may never converge to the global optimum. Though we may prevent the GA from getting stuck at a local optimum, the performance of the GA (in terms of the generations required for convergence) will certainly deteriorate.

To overcome the above-stated problem, we need to preserve "good" solutions of the population. This can be achieved by having lower values of p_c and p_m for high fitness solutions, and higher values of p_c and p_m for low fitness solutions. While the high fitness solutions aid in the convergence of the GA, the low fitness solutions prevent the GA from getting stuck at a local optimum. The value of p_m should depend not only on $f_{max} - \overline{f}$, but also on the fitness value f of the solution. Similarly, p_c should depend on the fitness values of both the parent solutions. The closer f is to f_{max}, the smaller p_m should be, i.e., p_m should vary directly as $f_{max} - f$. Similarly, p_c should vary directly as $f_{max} - f'$, where f' is the larger of the fitness values of the solutions to be crossed. The expressions for p_c and p_m now take the forms

$$p_c = k_1(f_{max} - f')/(f_{max} - \overline{f}), \qquad k_1 \leq 1.0 \qquad (3.1)$$

and

$$p_m = k_2(f_{max} - f)/(f_{max} - \overline{f}), \qquad k_2 \leq 1.0. \qquad (3.2)$$

(k_1 and k_2 have to be less than 1.0 to constrain p_c and p_m to the range 0.0 to 1.0).

Note that p_c and p_m are zero for the solution with the maximum fitness. Also, $p_c = k_1$ for a solution with $f' = \overline{f}$, and $p_m = k_2$ for a solution with $f = \overline{f}$. For solutions with subaverage fitness values, i.e., $f < \overline{f}$, p_c and p_m might assume values larger than 1.0. To prevent the overshooting of p_c and p_m beyond 1.0, we also have the following constraints:

$$p_c = k_3, \qquad f' \leq \overline{f} \qquad (3.3)$$

and

$$p_m = k_4, \quad f \leq \overline{f} \tag{3.4}$$

where $k_3, k_4 \leq 1.0$.

3.4.3 Practical Considerations and Choice of Values for k_1, k_2, k_3, and k_4

In the previous section, we saw that for a solution with the maximum fitness value, p_c and p_m are both zero. The best solution in a population is transferred undisrupted into the next generation. Together with the selection mechanism, this may lead to an exponential growth of the solution in the population and may cause premature convergence. To overcome the above-stated problem, we introduce a default mutation rate (of 0.005) for every solution in the AGA.

We now discuss the choice of values for k_1, k_2, k_3, and k_4. For convenience, the expressions for p_c and p_m are given below.

$$p_c = k_1(f_{max} - f')/(f_{max} - \overline{f}), \quad f' \geq \overline{f} \tag{3.5}$$

$$p_c = k_3, \quad f' < \overline{f} \tag{3.6}$$

and

$$p_m = k_2(f_{max} - f)/(f_{max} - \overline{f}), \quad f \geq \overline{f} \tag{3.7}$$

$$p_m = k_4, \quad f < \overline{f} \tag{3.8}$$

where $k_1, k_2, k_3, k_4 \leq 1.0$.

It has been well established in GA literature [5, 7] that moderately large values of p_c ($0.5 < p_c < 1.0$) and small values of p_m ($0.001 < p_m < 0.05$) are essential for the successful working of GAs. The moderately large values of p_c promote the extensive recombination of schemata, while small values of p_m are necessary to prevent the disruption of the solutions. These guidelines, however, are useful and relevant when the values of p_c and p_m do not vary.

One of the goals of our approach is to prevent the GA from getting stuck at a local optimum. To achieve this goal, we employ solutions with subaverage fitnesses to search the search space for the region containing the global optimum. Such solutions need to be completely disrupted, and for this purpose we use a value of 0.5 for k_4. Since solutions with a fitness value of $\overline{f}$ should also be disrupted completely, we assign a value of 0.5 to k_2 as well.

Based on similar reasoning, we assign k_1 and k_3 a value of 1.0. This ensures that all solutions with a fitness value less than or equal to $\overline{f}$ compulsorily undergo crossover. The probability of crossover decreases as the fitness value (maximum of the fitness values of the parent solutions) tends to f_{max}, and is 0.0 for solutions with a fitness value equal to f_{max}.

In the next section, we compare the AGA with previous approaches for employing adaptive operators in GAs.

3.5 Experiments and Results

In this section, we discuss the experiments that we have conducted to compare the performance of the AGA and SGA. For this purpose we have employed several multimodal test problems with varying complexities. The rest of this section is devoted to a discussion of the performance criteria for the GAs, the functions that are to be optimized, experiments, and the comparative results.

3.5.1 Performance Measures

As a measure of performance, we consider the average number of generations that the GA requires to generate a solution with a certain high fitness value (called the *threshold*). The average number of generations is obtained by performing the experiment repeatedly (in our case, 30 times) with different and randomly chosen initial populations.

Since the goal of our approach is to prevent the convergence of the GA to a local optimum, we also evaluate the performance of the GA in terms of the number of runs for which the GA gets stuck at a local optimum. When the GA fails to reach the global optimum after a sufficiently large number of generations, we conclude that it has gotten stuck at a local optimum.

3.5.2 Functions for Optimization

The choice of suitable functions to verify the performance of GAs is not an easy task. The nature of the optimization function varies a lot from application to application, in terms of the number of local optima, the rate of variation of the objective function, etc. In this research, we have used several multimodal functions with varying complexities. They are the following:

DeJong's f5: This is a spiky function (also known as Shekel's foxholes) with 25 sharp spikes of varying heights. The function has two variables and the solution is encoded using 34 bits. The task of the GA is to locate the highest peak. The expression for f5 is as follows.

$$f5 = 0.002 + \sum_{j=1}^{25} \frac{1}{j + \sum_{i=1}^{2} (x_i - a_{ij})^6}$$

f6: This is a rapidly varying multimodal function of two variables and is symmetric about the origin with the height of the barrier between adjacent minima increasing as the global optimum is approached. The variables are encoded using 22 bits each and assume values in the range $(-100.0, 100.0)$. f6 has been employed earlier [18] for comparative studies, where it is referred to as the "sine envelope sin wave function". The expression for f6 is as follows:

$$f6 = 0.5 + \frac{sin^2\sqrt{x_1^2 + x_2^2} - 0.5}{\left[1.0 + 0.001(x_1^2 + x_2^2)\right]^2}$$

f7[1]: This function is also similar to f6, but has the barrier height between adjacent minima approaching zero as the global optimum is approached:

$$f7 = (x_1^2 + x_2^2)^{0.25}\left[sin^2(50(x_1^2 + x_2^2)^{0.1}) + 1.0\right]$$

Order-3 deceptive: GA-deceptive functions are being used extensively to evaluate the performance of GAs. The order-3 deceptive function depends on three binary bits as shown in Table 3.1. Optimizing the 3-bit deceptive function is a trivial

Table 3.1 A 3-Bit Deceptive Function

Binary code	Function value
000	28
001	26
010	22
011	0
100	14
101	0
110	0
111	30

exercise. The actual function to be optimized by the GAs is the sum of five such independent functions. The solution string is obtained by concatenating five of the 3-bit codes. We used only five subfunctions in our function, mainly to enable the GAs to converge to the optimum with small populations (population size =

---

[1]The decoded value of each variable of the functions f5, f6, and f7 has been shifted by 10% to the left, and wrapped around the upper limit in case the value is less than the lower limit. This has been done to shift the optimal solutions away from Hamming cliffs (see Reference [4]).

100). For a further description of deceptive functions the reader is referred to Reference [8].

TSP: The traveling salesman problem (TSP) involves finding the shortest Hamiltonian cycle in a complete graph of n nodes. The Euclidean distance between any two nodes is computed from their coordinates. An instance of the TSP is specified by n, the number of cities, and the coordinates of the n cities (nodes). In our implementations, we have employed the *order crossover operator* [7, 16], and a mutation operator that swaps the positions of two randomly chosen cities. p_c and p_m determine the probability with which the operators are employed. We have chosen the 30-city and 105-city problems (see Reference [22] for coordinates of cities) for comparing the performance of the SGA and the AGA.

Neural networks: The underlying optimization problem in feed-forward neural networks is that of identifying a set of interconnection weights, such that a mean square error defined between a set of output patterns and training patterns is minimized. Each neuron i may be associated with

- An output value O_i

- A set of k input values I_{ji} $1 \leq j \leq k$

- A threshold value T_i

- A set of interconnection weights w_{ji} $1 \leq j \leq k$

- An *activation value* $A_i = \sum_{j=1}^{k} w_{ji} I_{ji} - T_i$

The output value of each neuron is typically a nonlinear function of the activation value. In a feed-forward network, the neurons are organized into layers (input, output, and hidden), with the inputs of each neuron connected to the outputs of the neurons of the previous layer. The input patterns are applied to the input layer, and the training pattern is compared with the outputs of neurons in the output layer. The mean square error for a given set of weights is evaluated as

$$MSE = \frac{1}{pN_o} \sum_{i=1}^{p} \sum_{j=1}^{N_o} (O_{ij} - O'_{ij})^2 \qquad (3.9)$$

where

MSE:	the mean square error
p:	number of input patterns
N_o:	number of output neurons
O_{ij}:	output value of the jth neuron for the ith input pattern
O'_{ij}:	training value of jth neuron for the ith input pattern

In our implementation, the output function f is sigmoidal: $f_i = (1 + exp^{-10A_i})^{-1}$. We also use binary inputs and train the network to generate binary outputs. Further, $T_{ij} = 0.1$ for a binary 0 and $T_{ij} = 0.9$ for binary 1. w_{ij} and T_j assume values in the range -1.0 to $+1.0$. Each weight is encoded using 8 bits, and the string is formed by concatenating the binary codes for all the weights and threshold values.

We consider three mapping problems:

- **XOR**: 2 inputs, 1 output , 5 neurons, 9 weights, 4 input patterns; the output value is the exclusive OR of the input bits.

- **4-bit parity**: 4 inputs, 1 output, 9 neurons, 25 weights, 16 input patterns; the output value is 1 if there are an odd number of 1s among the inputs.

- **Decoder encoder**: 10 inputs, 10 outputs, 25 neurons, 115 weights, 10 input patterns (each having all 0s and a 1 at one of the ten inputs); output pattern is the same as the input pattern.

Test generation problem: The primary task of *test generation* for digital logic circuits is to generate input vectors of logical 0s and 1s that can check for possible faults in the circuit by producing observable faulty response at the primary outputs of the circuit. The problem of generating a test for a given fault has been proved to be NP-complete [14]. In generating tests, it is desirable to detect close to 100% of all the possible faults in the circuit. Test generation as a candidate optimization problem for GAs may be characterized as follows:

- Faults are modeled as being stuck-at-0 or stuck-at-1.

- A test for a fault should (1) generate a logic value at the fault site that is different from the stuck-at value of the fault, (2) should be able to propagate the fault effect to one of the primary outputs.

- Fault simulation approach to test generation: input vectors are generated randomly, and then through logic simulation, the faults that the vector detects are identified as being detected.

- Random test generation may be improved by using a search based on a cost associated with each input vector.

- Distance cost function: $C_v = \sum_{i \in F} L_m - L_{vi}$
  where

C_v:	cost associated with a vector v
F:	set of undetected faults
L_m:	maximum number of gate levels in the circuit
L_{vi}:	level to which the fault effect of i has been propagated by vector v

The cost C_v is minimum (locally) when a given input vector is a test for a certain fault. It should be noted that the cost function changes as faults are detected and removed from the list of undetected faults. The task for the GA is to minimize the cost C_v. Test circuits for experiments have been chosen from the ISCAS-85 benchmarks [15].

3.5.3   Experimental Results

Except for the TSPs, in all our experiments we have used a population size of 100 for the GAs. "Scaling" of fitness values and the stochastic remainder technique (see Reference [7]) for "selection" have been used in the GAs. All parameters have been encoded using a fixed point encoding scheme.

For the SGA, we have used values of $p_c = 0.65$ and $p_m = 0.008$.

For the AGA, p_c and p_m are determined according to Expressions 3.5 to 3.8 given in Section 3.4.3.

Table 3.2   Comparison of Performance of AGA and SGA

Function	String length	Generations SGA	AGA	Stuck SGA	AGA	Thresh	Max generations
XOR	72	61.20	36.73	10	0	0.999	100
4-bit parity	200	399.33	93.43	18	0	0.999	500
Deceptive encoding	920	456.43	71.70	26	0	0.99	500
f5	34	64.06	36.63	7	0	1.00	100
f6	44	173.9	106.56	23	6	0.999	200
f7	44	419.90	220.61	21	5	0.995	500
Order-3 deceptive	15	70.32	105.33	8	9	1.00	200

Table 3.3   Performance of AGA and SGA for the TSP

Cities	Avg tour length SGA	AGA	Optimum tour located SGA	AGA	Max generations	Population size
30 (424.0)	442.1	430.2	0	7	100	1,000
105 (14,383)	16,344.3	14,801.4	0	4	500	2,000

The experimental results are presented in Tables 3.2 to 3.4. Table 3.2 gives the average number of generations required by each GA for attaining a solution with a fitness value equal to the threshold value "Thresh"[2]. Also tabulated is the

---

[2]We are not measuring population convergence based on the mean convergence of each bit, since the AGA never converges in the above sense due to the high disruption rates of low fitness solutions.

Table 3.4 Performance of AGA and SGA for the Test Generation Problem

Circuit	SGA (generations)	AGA (generations)	Fault coverage	String length
c432	102.10	10.73	99.23 %	36
c499	10.91	10.50	98.94 %	41
c880	155.23	37.33	100.00%	60
c1355	35.26	31.70	99.49%	41
c1908	122.13	57.93	99.52 %	33
c3540	155.43	73.66	96.00%	50
c5315	53.33	21.56	98.89%	178

number of instances (out of 30 trials) for which the GAs have gotten stuck at a local optimum. The maximum number of generations that the GAs were executed for, and the string length are also indicated for each of the problems.

The AGA outperforms the SGA for all the problems except order-3 deceptive. For the three neural network problems, the AGA has located the optimal solution in every trial, while the performance of the SGA has been poor.

For the TSPs we have used populations of 1000 and 2000 for the 30- and 105-city problems, respectively. The number of function evaluations have been 100,000 and 1,000,000, respectively. Results have been obtained for ten different trials. For both the problems, the SGA was not able to locate the optimal tour even on one occasion, while AGA's performance has been significantly better, both in terms of the average tour length and the number of instances when the optimal tour was located.

Table 3.4 compares the performance of the AGA and the SGA for the test generation problem. The numeral in the circuit name indicates the gate count of the circuit. Once again, the superior performance of AGA is clear. For c432, the SGA requires almost ten times the number of generations that the AGA needs to locate all detectable faults. Only for c499, the SGA has come close to performing as well as the AGA. The results are averages over 30 different trials for each circuit. It may be noted that the complexity of test generation is not directly dependent on the circuit size, but is controlled by several other factors such as the fan-ins and fan-outs of gates, the number of levels in the circuit, etc.

3.5.4 When Does the AGA Perform Well?

The optimization problems considered above span a range of complexities, string lengths, and problem domains. In general, the performance of the AGA has been significantly superior to that of the SGA, while in specific instances such as the order-3 deceptive problem and for c499 in the test generation problem, the SGA has performed as well as the AGA or better. The experimental results also point out that the relative performance of the AGA as compared to that of the SGA varies considerably from problem to problem. All the problems that we have considered have some epistaticity present; however, the extent of epistaticity varies considerably. For instance, in the neural network problems, the high epistaticity is brought about by the fitness being a complex nonlinear function of the weights, while in the order-3 deceptive problem, the epistaticity is relatively lower with the fitness contribution due to a bit being affected only by two other bits.

A different aspect of multimodal function optimization is the sensitivity of the optimization technique to the "multimodality" of the problem, i.e., how the performance varies as the number of local optima in the search space varies. It may be observed from Table 3.2 that the relative performance of the AGA with respect to that of the SGA is better for f6 and f7 than for f5. Although the evidence is not conclusive, it appears that the AGA performs relatively better than the SGA when the number of local optima in the search space is large.

To better understand the circumstances under which AGA performs better than the SGA, we have conducted two sets of experiments where we have methodically varied the epistacity in the problem and the number of local optima in the search space. For purposes of convenience, we have chosen the following objective function for these experiments,

$$f8 = \sum_{i=1}^{P} \left| \frac{sin(\pi k x_i)}{(\pi k x_i)} \right| \quad -0.5 \leq x_i \leq 0.5$$

where P gives the number of variables.
The function has one global optimum and P^k local optima for odd values of k.

To characterize the effect of varying the number of local optima in the search space, we have varied k for a fixed value of $P = 5$. Each variable x_i is encoded using 10 bits. The experimental results are presented in Table 3.5. Table 3.5 confirms our earlier observation that, with increasing number of local optima the performance of AGA improves steadily over that of the SGA. For $k = 1$, the function is unimodal and the SGA outperforms the AGA.

Next we consider the effects of varying the epistaticity of the function. We consider strings of length 40, $k = 5$, and we vary the number of parameters P. Correspondingly, the number of bits required for encoding each variable also changes. The epistacity increases as P decreases, since the fitness due to a single variable x_i depends on the interactions of $40/P$ bits. From Table 3.6 it is clear that the relative performance of the AGA with respect to the SGA deteriorates as the epistaticity decreases.

3.5.5   Sensitivity of AGA to k_1 and k_2

We have already pointed out in Section 3.2 that p_c and p_m critically control the performance of the GA. One of the goals of having adaptive mutation and crossover

Table 3.5 Effect of
"Multimodality" on the
Performance of the AGA and
SGA

	Generations		Stuck	
k	SGA	AGA	SGA	AGA
1	24.10	27.73	0	0
3	75.56	70.91	4	0
5	78.96	71.93	12	3
7	82.76	72.70	13	4

Table 3.6 Effect of Epistaticity on the Performance of the AGA and SGA

		Generations		Stuck	
Code Length	Variables	SGA	AGA	SGA	AGA
20	2	92.50	86.13	11	7
10	4	71.73	64.36	5	4
8	5	59.33	52.40	3	3
5	8	35.10	48.86	1	2
4	10	27.73	42.66	0	0

Table 3.7 Effect of $k1$ on AGA Performance: Average Number of Generations for Convergence and Number of Instances When AGA Gets Stuck at a Local Optimum

Function	$k1 = 0.2$	0.4	0.6	0.8	1.0
f5	47.76 (6)	40.20 (3)	33.46 (3)	41.33 (1)	36.63 (0)
XOR	45.70 (2)	37.36 (0)	42.30 (0)	38.56 (0)	36.73 (0)
Order-3 deceptive	143.13 (11)	125.66 (9)	109.7 (10)	121.66 (13)	105.33 (9)

is to ease the user's burden of specifying p_m and p_c. However, our method has introduced new parameters k_1 and k_2 for controlling the adaptive nature of p_m and p_c. To evaluate the effect of k_1 and k_2 on the performance of the AGA, we have monitored the performance of the AGA for varying values of k_1 (0.2 to 1.0) and k_2 (0.1 to 0.5). The experimental results are presented in Tables 3.7 and 3.8.

On analyzing the results presented in Table 3.8 for different values of k_2, we notice no dramatic difference in the performance of the AGA in terms of the average number of generations required for convergence. The fact that the performance of the GA hardly varies with the value of k_2 shows that the AGA is not sensitive to

Table 3.8   Effect of $k2$ on AGA Performance: Average Number of Generations for Convergence and Number of Instances When AGA Gets Stuck at a Local Optimum

Function	$k2 = 0.1$	0.2	0.3	0.4	0.5
f5	60.20 (2)	48.26 (1)	65.33 (1)	44.23 (2)	36.63 (0)
XOR	52.26 (0)	51.66 (0)	49.73 (0)	44.60 (0)	36.73 (0)
Order-3 deceptive	122.30 (15)	117.43 (12)	98.66 (8)	96.53 (6)	105.33 (9)

the external parameter k_2—one of the goals of our research. However, the AGA gets stuck at local optima fewer times for higher values of k_2 than for lower values of k_2. The results justify our choice of $k_2 = 0.5$ for the AGA.

Table 3.7 demonstrates the steady improvement in performance of AGA as k_1 is increased. This may be expected since a large value of k_1 maximizes the recombination of schemata, while the best schemata are yet retained due to the adaptation policy.

3.6 Conclusions

Recent research on GAs has witnessed the emergence of new trends that break the traditional mold of "neat" GAs that are characterized by static crossover and mutation rates, fixed length encodings of solutions, and populations of fixed size. Dynamic variation of control parameters based on an adaptive policy has been employed in several implementations to improve the efficacy of GA search.

In this chapter we have surveyed some adaptive strategies for varying the control parameters of the GA. Specifically, we delve into the details of the AGA, which reflect our own work in this direction. We adopt a "messy" approach to determine p_c and p_m, the probabilities of crossover and mutation. The approach is different from the previous techniques for adapting operator probabilities, as p_c and p_m are not predefined—they are determined adaptively for each solution of the population. The values of p_c and p_m range from 0.0 to 1.0 and 0.0 to 0.5, respectively. It might appear that the low values of p_c and the high values of p_m might either lead to premature convergence of the GA or transform the GA into a random search. However, it is the manner in which p_c and p_m are adapted to the fitness values of the solutions that not only improves the convergence rate of the GA, but also prevents the GA from getting stuck at a local optimum. In the adaptive GA, low values of p_c and p_m are assigned to high fitness solutions, while low fitness solutions have very high values of p_c and p_m. The best solution of every population is "protected", i.e., it is not subjected to crossover and receives only a minimal amount of mutation. On the other hand, all solutions with a fitness value less than the average fitness value of the population have $p_m = 0.5$. This means that all subaverage solutions are completely disrupted and totally new solutions are created. The GA can, thus, rarely get stuck at a local optimum.

We have conducted extensive experiments on a wide range of problems including TSPs, neural network weight-optimization problems, and generation of test vectors for VLSI circuits. In most cases, the AGA has outperformed the SGA significantly. Specifically, we have observed that, for problems that are highly epistatic and multimodal, the AGA performs very well.

In this work, we have chosen one particular way of adapting p_c and p_m based on the various fitnesses of the population. The results are encouraging, and future work should be directed at developing other such adaptive models for the probabilities of crossover and mutation. A similar dynamic model for varying the population size in relation to the fitnesses of the population is certainly worth investigating.

References

[1] Davis, L., Ed., *Genetic Algorithms and Simulated Annealing*, Pitman Publishing, London, 1987.

[2] Davis, L., Adapting operator probabilities in genetic algorithms, in Proc. 3rd Int. Conf. of Genetic Algorithms, Morgan Kaufmann, San Mateo, CA, 1989, 61.

[3] Davis, L., Ed., *Handbook of Genetic Algorithms*, Van Nostrand Reinhold, New York, Morgan Kaufmann, San Mateo, CA, 1991.

[4] Davis, L., Bit climbing, representational bias, and test suite design, in Proc. 4th Int. Conf. of Genetic Algorithms, 1991, 18.

[5] DeJong, K.A., An Analysis of the Behaviour of a Class of Genetic Adaptive Systems, Ph.D. dissertation, University of Michigan, Ann Arbor, 1975.

[6] DeJong, K.A., Genetic algorithms: a 10-year perspective, in Proc. Int. Conf. of Genetic Algorithms and Their Applications, Greffenstette, J., Ed., Pittsburgh, July 24–26, 1985, 169.

[7] Goldberg, D.E., *Genetic Algorithms in Search, Optimization and Machine Learning*, Addison-Wesley, Reading, MA, 1989.

[8] Goldberg, D.E., Genetic algorithms and Walsh functions: II. Deception and its analysis, *Complex Syst.*, 3, 153, 1989.

[9] Goldberg, D.E. et al., Messy genetic algorithms: motivation, analysis and first results, *Complex Syst.*, 3, 493, 1989.

[10] Goldberg, D.E. et al., Messy genetic algorithms revisited: studies in mixed size and scale, *Complex Syst.*, 4, 415, 1990.

[11] Grefenstette, J.J., Optimization of control parameters for genetic algorithms, *IEEE Trans. Syst., Man Cybern.*, SMC-16(1), 122, 1986.

[12] Holland, J.H., *Adaptation in Natural and Artificial Systems*, University of Michigan Press, Ann Arbor, 1975.

[13] Cobb, H.C., An Investigation into the Use of Hypermutation as an Adaptive Operator in Genetic Algorithms Having Continuous Time-Dependent Non-stationary Environments, AIC-90-001, NRL Report 6760, Naval Research Laboratory, Washington, D.C., 1990.

[14] Ibarra, O.H. and Sahni, S.K., Polynomially complete fault detection problems, *IEEE Trans. Comput.*, C-24, 242, 1975.

[15] Special session: recent algorithms for gate-level ATPG with fault simulation and their performance assessment, *Proc. 1985 IEEE Int. Symp. Circuits Syst. (ISCAS)*, p. 663, June 1985.

[16] Oliver, I.M., Smith, D.J., and Holland, J.R.C., A study of permutation crossover operators on the travelling salesman problem, in Proc. 2nd Int. Conf. of Genetic Algorithms, Lawrence Erlbaum, 1987, 224.

[17] Smith, R.E., Adaptively Resizing Populations: An Algorithm and Analysis, TCGA Report No. 93001, Clearinghouse for Genetic Algorithms, Department of Engineering Mechanics, University of Alabama, Tuscaloosa, 1993.

[18] Schaffer, J.D. et al., A study of control parameters affecting online performance of genetic algorithms for function optimization, in Proc. 3rd Int. Conf. of Genetic Algorithms, Morgan Kaufmann, San Mateo, CA, 1989, 51.

[19] Srinivas, M. and Patnaik, L.M., Adaptive probabilities of crossover and mutation in genetic algorithms, *IEEE Trans. Syst. Man Cybern.*, 24, 4, 17–26, April 1994.

[20] Srinivas, M. and Patnaik, L.M., Genetic algorithms: a survey, *IEEE Comput.*, p. 17, June 1994.

[21] Spears, W.M. and DeJong, K.A., An analysis of multipoint crossover, in Proc. 1990 Workshop of the Foundations of Genetic Algorithms, Rawlins, G., Ed., Morgan Kaufmann, San Mateo, CA, 1991, 301.

[22] Whitley, D. and Starkweather, D., Genitor-II: a distributed genetic algorithm, *J. Exp. Theor. Artif. Int.*, 2, 189, 1990.

4

An Empirical Evaluation of Genetic Algorithms on Noisy Objective Functions

Keith Mathias, Darrell Whitley, Anthony Kusuma, and Christof Stork

Abstract   Genetic algorithms have particular potential as a tool for optimization when the evaluation function is noisy. Several types of genetic algorithms are compared against a mutation-driven stochastic hill-climbing algorithm on a standard set of benchmark functions which have had Gaussian noise added to them. Different criteria for judging the effectiveness of the search are also considered. The genetic algorithms used in these comparisons include an elitist simple genetic algorithm, the CHC adaptive search algorithm, and the delta coding genetic algorithm. Finally, several hybrid genetic algorithms are described and compared on a very large and noisy seismic data imaging problem.

4.1   Introduction

Fitzpatrick and Grefenstette have provided empirical and analytic evidence suggesting that genetic algorithms exhibit a certain tolerance for noise [6]. In this chapter the performance of several genetic algorithms are compared against that of a mutation-driven stochastic hill-climber on a set of relatively difficult benchmark functions where Gaussian noise has been injected into the evaluation function. The genetic algorithms used in these comparisons include an elitist simple genetic algorithm, the CHC adaptive search algorithm, and the delta coding genetic algorithm. These comparisons are accomplished using two different optimality metrics. In one case, we look at performance with respect to finding a noisy solution within a known range of the actual solution; in the other case, we look at each algorithm's ability to locate the true optimum. However, in both sets of comparisons the genetic algorithm only receives the noisy function evaluation value as feedback.

Additional performance comparisons involving a version of the delta coding algorithm combined with a local search algorithm were carried out for a seismic data interpretation problem known to have a noisy objective function. Because the delta coding algorithm is hybridized with a local steepest ascent algorithm, the comparisons for this application do not include the same set of algorithms used in the comparisons on the noisy benchmark functions. Instead the hybridized delta coding algorithm is compared to other hybridized algorithms that combine local and genetic search. These algorithms have been specifically designed for the seismic data interpretation problem.

Seismic reflection surveys are used to construct subsurface images of geologic beds. The images can be distorted by high levels of *ambient noise* and by effects associated with surface materials through which the seismic signal travels. However, these images can be corrected by application of "static" corrections or shifts. This problem is characterized by highly nonlinear interactions and large search spaces (e.g., 2^{6000} points is not uncommon) with many local optima [16]. The evaluation function is also noisy in the sense that the true evaluation function involves computing cross-correlations between numerous seismic signals (e.g., 500) at each of 500 to 1000 time steps. To reduce the cost of evaluation, the cross-correlations are sampled at a rate of 10%, thus making the evaluation function noisy and approximate.

4.2   Background

In initial studies involving noisy environments [10] the delta coding genetic algorithm [18] has performed well. Delta coding begins by running a genetic algorithm while monitoring the diversity of the population. Our version of delta coding uses GENITOR [17] as the genetic search component. The GENITOR algorithm ranks the population; two offspring are then selected with a random linear bias toward the best members of the population. One offspring is produced, which replaces the worst member of the population. The new offspring is then assigned a rank in the population according to fitness. Note that the worst ranked member of the population is continually replaced; hence this position is really a kind of buffer or place holder, especially when new offspring have fitness values poorer than the members of the current population and thus are immediately replaced after the next recombination. We will refer to the *persisting population* as the normal population excluding this worst string buffer position.

One of the key components of the delta coding algorithm is a restart mechanism. Genetic search is stopped as soon as the population shows signs of losing diversity. Population diversity is monitored by testing the Hamming distance between the best and worst individuals in the *persisting population*. If it is greater than one, genetic search continues; otherwise, genetic search is temporarily suspended.

After a delta coding run is halted the search is restarted. The best solution parameters are saved as the *interim solution*. The population is then randomly re-initialized and the genetic algorithm is restarted, again monitoring the population diversity. Genetic search proceeds as normal except that the parameter substrings are decoded such that they represent a distance or *delta value* ($\pm\delta$) away from the interim solution parameters. Thus, these delta values are added to or subtracted from the interim solution parameters. The resulting parameters can then be evaluated using the original fitness function. This remaps the search space such that the interim solution is located at the origin of the hypercube. Additionally, the relationships between strings in Hamming space in the previous mapping are rearranged with respect to the schemata competitions that occur during genetic search. When the population diversity has been adequately exploited again the search is suspended and a new set of interim solution parameters are saved. This cycle is repeated until the optimum solution is located or until some specified amount of work has been expended.

Delta coding also includes mechanisms for reducing and enlarging the size of the subpartition that is currently being searched. Each time a new interim solution is saved the number of bits used to represent each of the function parameters is altered. If the new interim solution parameters are different than those used in the previous iteration, then the parameter bit representations are decreased by one bit each. This allows the search to focus on particularly promising subpartitions of the search space. However, if the algorithm converges to the same solution (i.e., the delta values are all zero) then the parameter bit representations are increased by one. This keeps the algorithm from getting stuck in suboptimal subpartitions of the search space.

An earlier study indicated that delta coding performed better than the other algorithms considered in this study on DeJong's noisy test function (F4) when the performance of the algorithm is measured with respect to finding a noisy solution within a specific range of the known optimum. These results are reprinted in Table 4.1 [10]. Each algorithm was tested 30 independent times allowing a maximum of 100,000 trials for each run. The stopping criteria for these tests were to find a string with an evaluation of -2.5 returned from the noisy evaluation function (except for the ESGA tests, where a value of -2.0 was used in order to be consistent with most other previous studies). The -2.5 stopping criterion is based on the fact that the optimum is 0.0 and the added noise has a standard deviation (σ) of 1.0. No Gray coding was used to transform the binary parameter representation for the problem to avoid compromising the difficulty of the problem [11].

Table 4.1 Performance Comparisons for DeJong's F4 Noisy Test Function

Algorithm	Stochastic hill-climber	ESGA	GENITOR	CHC	Delta coding
Percent solved	20%	90%	100%	100%	100%
Average trials	44,967	28,677	17,367	17,017	5,027
Average best	−1.73	−2.08			
Population size	1	10	100	50	25
Crossover rate		0.70			
Linear selection bias			1.25		2.0
Mutation rate	0.02	0.005	0.01	35%	0.03

4.3 Empirical Benchmarks

To further evaluate the effectiveness of genetic algorithms in noisy environments we tested three genetic algorithms and a stochastic hill-climber on four benchmark test functions where Gaussian noise was artificially injected into the objective function. A binary representation is used. The functions selected for these tests included three of the larger benchmark functions used in the genetic algorithm community: the Rastrigin (F6), Schwefel (F7), and Griewank (F8) functions [13]. Gaussian noise with a mean of zero and a standard deviation of 1.0 was added to the function evaluation value for each of the functions, as in DeJong's F4 function. DeJong's F4 function was included in these tests as well. The objective of adding a Gaussian noise component to existing functions was to evaluate the performance of the algorithms for a set of functions where the performance results without the noise were already known. The equations for these test functions with the Gaussian noise component added are

$$F4: \quad f(x_i \mid_{i=1,30}) = \left[\sum_{i=1}^{30} i x_i^4 \right] + Gauss(0, 1) \qquad x_i \in [-1.28, 1.27]$$

$$F6: \quad f(x_i \mid_{i=1,20}) = \left\{ (200) + \left[\sum_{i=1}^{20} \left(x_i^2 - 10 \cos(2\pi x_i) \right) \right] \right\}$$

$$+ Gauss(0, 1) \qquad x_i \in [-5.12, 5.11]$$

$$F7: \quad f(x_i \mid_{i=1,10}) = \left[\sum_{i=1}^{10} -x_i \sin(\sqrt{|x_i|}) \right] + Gauss(0, 1) \qquad x_i \in [-512, 511]$$

$$F8: \quad f(x_i \mid_{i=1,10}) = \left[1 + \sum_{i=1}^{10} \frac{x_i^2}{4000} - \prod_{i=1}^{10} (\cos(x_i/\sqrt{i})) \right]$$

$$+ Gauss(0, 1) \qquad x_i \in [-512, 511]$$

4.3.1 Algorithm Descriptions

The mutation-driven stochastic hill-climbing algorithm used in these comparisons begins by randomly generating a single binary string. Search is performed by applying bit mutation to that single string. The changes were kept only when the fitness of the resulting offspring was better than or equal to the fitness of the string before mutation. This process was repeated until the optimal solution was found or until some maximum number of trials were executed. This results in a simple stochastic hill-climbing algorithm. This algorithm was included in these tests as a benchmark for understanding the difficulty of the problems [4].

We also tested an elitist simple genetic algorithm (ESGA) [7] where the best individual in the population is always copied into the next generation. Selection in our ESGA is performed using Baker's stochastic universal sampling algorithm [1]. Two-point reduced surrogate crossover was used for recombination [2] and was applied according to the probability, P_c. The probability of mutation, P_m, is the probability that a bit is flipped.

The CHC adaptive search algorithm [5] was included in these comparisons because of the algorithm's outstanding performance in other studies [5, 10]. The CHC algorithm is a generational genetic search model that employs a cross-generational selection/competition mechanism. Stings are randomly and uniformly chosen for recombination from the parent population. Offspring are held in a temporary population. Then the best N strings from the parent and offspring populations are selected for the next generation, where N is the population size ($N = 50$ for all tests performed here). If no offspring can be inserted into the population for the next generation CHC uses *cataclysmic mutation* [5] to restart the search.

Cataclysmic mutation keeps one copy of the best individual in the population. Then that string is used as a template to reinitialize the remainder of the population. These strings are formed by repeatedly mutating some percentage of bits in the template string and copying the result into the new population. A cataclysmic mutation rate of 35% was used for all of these experiments.

CHC also employs a unique crossover operator (HUX) to facilitate "heterogeneous recombination" and prevent "incest" [5]. The HUX operator compares potential mates and if the number of differing bits exceeds some threshold, uniform crossover is randomly applied to half of those positions that differ. HUX is a very disruptive crossover operator designed to scatter offspring to random points that are equidistant between parents in hyperspace.

4.4 Performance Comparisons Using Noisy Fitness Values to Approximate Optimality

Progress here was measured with respect to the noisy function evaluation value. Search was terminated when the noisy function evaluation value returned to the algorithm was 2.5σ (standard deviations) below the value of the true optimal solution. Thus, if the optimal solution for a test function was 0.0, then the stopping criteria would be met when the noisy function evaluation returned a value less than or equal to -2.50. Use of the noisy function evaluation value in this manner allows a range of solutions (conditional on the noise value) to be accepted as a final solution.

Previously published algorithm comparisons have used a wide variety of metrics to track the performance of these algorithms on DeJong's F4 function. Even the uses of the noisy function evaluation criteria vary widely in research literature. Additionally, the comparisons in Section 4.2 do not impose the same criteria on all of the algorithms tested. For example, the progress of the ESGA was measured using a noisy function evaluation value of -2.0, while the other algorithms were compared using a value of -2.5 to indicate optimality. For the comparisons here all algorithms are measured using the same criteria (-2.5σ).

Previous genetic algorithm performance comparisons using functions included in this test suite report the use of Gray coding [5, 8, 10, 15]. However, Gray coding was not employed in these tests so that the integrity of the functions might not be compromised [11]. Previous studies also indicated that the use of Gray coding did not significantly influence the performance of an ESGA on DeJong's F4 test function [3].

For these performance comparisons the various algorithm parameters were set to match those found in previous studies [10] and then tuned by judgment so as to give the best performance possible for each algorithm. This reduces the chance of biasing the comparisons by selecting a single set of parameters that enhance the performance of one algorithm while diminishing the performance of the others. The best possible performance for these tests was defined as: (1) the most number of runs matching the *noisy* stopping criteria value for the function while performing the fewest number of recombinations; or (2) when the stopping criteria are not always satisfied, the smallest average solution in the maximum number of trials allowed for the tests.

4.4.1 Empirical Results and Analysis

The performance results for each of the four algorithms tested on these functions with Gaussian noise artificially injected into the evaluation function are presented in Table 4.2. Percent solved indicates the percentage of 30 independent runs where the noisy function evaluation value returned to the genetic algorithm reached 2.5σ

Table 4.2 Algorithm Performance Comparisons on Noisy Functions Using Noisy Evaluation Function Value Optimality Approximation Metric

Algorithm	Problem (no. params)	F4 (30)	F6 (20)	F7 (10)	F8 (10)
	Total bits	240	200	100	100
	Maximum trials	100,000	500,000	200,000	500,000
Stochastic hill-climber	Percent solved	27%	0%	0%	10%
	Average trials	54,539			311,284
	σ trials	29,983			165,239
	Average best (noisy)	−1.96	16.70	−4,052.29	−1.92
	σ	0.62	4.50	111.73	0.40
	Average best (true)	1.72	19.65	−4,050.11	1.79
	σ	0.76	4.57	111.49	0.27
	Mutation rate	0.02	0.04	0.07	0.07
ESGA	Percent solved	40%	0%	0%	13%
	Average trials	37,486			273,115
	σ trials	28,260			117,362
	Average best (noisy)	−2.32	14.36	−4,184.79	−2.24
	σ best	0.48	3.53	21.02	0.33
	Average best (true)	1.46	17.16	−4,182.10	1.70
	σ	0.64	3.61	20.97	0.31
	Population size	10	50	50	50
	Crossover rate	0.70	0.80	0.60	0.90
	Mutation rate	0.005	0.01	0.01	0.005
CHC	Percent solved	100%	53%	100%	97%
	Average trials	19,493	321,322	25,243	132,278
	σ trials	8,694	111,752	8,877	73,185
	Average best (noisy)	−2.72	−0.21	−4,192.51	−2.53
	σ		3.01		0.12
	Average best (true)	0.62	2.01	−4,189.19	1.33
	σ	0.27	1.62	0.29	1.13
	Population size	50	50	50	50
	Cataclysmic mutation	35%	35%	35%	35%
δ coding	Percent solved	100%	73%	83%	100%
	Average trials	5,989	294,902	34,631	80,670
	σ trials	4,852	70,943	6,927	52,035
	Average best (noisy)	−2.71	−2.58	−4,158.9	−2.63
	σ best		0.27	83.37	
	Average best (true)	0.74	1.09	−4,155.70	1.32
	σ	0.21	0.43	83.24	0.12
	Population size	25	800	100	200
	Linear selection bias	2.00	1.90	1.90	2.00
	Mutation rate	0.03	0.01	0.02	0.005

below the known optimal value of the function without noise. Average trials indicates the number of recombinations performed by the algorithm for those runs that found the optimal solution. The value for noisy average best indicates the average best solution found with respect to the noisy function evaluation value returned to the genetic algorithm (over 30 runs) after the maximum number of trials. The noisy σ is the standard deviation associated with the noisy average best value. The true average best is the average of the *true* function evaluation value without noise for the same 30 strings evaluated for the noisy average best value.

The results presented in Table 4.2 indicate that the addition of Gaussian noise to these benchmark functions provides a challenging noisy test suite. The results for the ESGA on DeJong's F4 function show that the function is much harder when the noisy function evaluation value must reach -2.5σ instead of -2.0σ, as used in previous tests (see Table 4.1).

The mutation-driven stochastic hill-climber performs worse than any of the genetic algorithms tested here on all of the functions in this test suite. This indicates that these functions are not easily hill-climbed and may also imply that the population-based genetic search algorithms are able to filter noise more effectively than the stochastic hill-climber.

The results also indicate that the addition of noise and the use of the noisy function evaluation value to measure the performance of the algorithms make the Griewank (F8) function much easier to optimize than the actual function without noise. (This appears to be due to the fact that there are now multiple solutions that match the stopping criteria; it may also be significant to note that there are many near-optimal solutions to the one-dimensional version of the Griewank function that surround the global optimum.) The CHC algorithm was only able to locate the optimal solution 23% of the time for the normal F8 function without noise; CHC found a solution satisfying the stopping criteria on 97% of the runs on the noisy version of the function. Likewise, the stochastic hill-climber was not able to solve the original Griewank function without noise even a single time but was able to find a satisfactory approximation for the noisy version of the function 10% of the time. It should be noted, however, that all of the average true fitnesses recorded for the F6, F7, and F8 functions are much worse than in the original versions of the problems. Thus, the actual optimization of the underlying versions of these functions without noise is much less successful when noise is injected into the function evaluation.

The noisy versions of the F6, F7, and F8 functions seem to be much harder for the delta coding algorithm than the original version of the functions. In fact, delta coding was the only algorithm to solve the original F6 and F8 functions to optimality in previous tests when Gray coding was not used and noise was not injected into the fitness functions. Delta coding was also able to consistently locate the optimal solution for the original version of the F7 function, albeit somewhat slower than the CHC algorithm. However, even when the noisy evaluation metric is used the performance of delta coding is much worse on the F6, F7, and F8 functions. Delta coding is not even able to consistently locate the optimal solution for the noisy versions of the F6 and F7 functions. Even so, delta coding still performs better than the other algorithms tested here on the F4, F6, and F8 functions with respect to the percent of the runs that were able to locate a noisy solution that satisfied the stopping criteria.

The CHC algorithm does not perform quite as well as delta coding on the noisy F4, F6, and F8 test functions. However, the CHC algorithm is able to approximate optimal solutions to the noisy versions of the Rastrigin and Griewank functions with more reliability than it is able to find the true optimal solutions of the original versions of the functions. The results in a previous study with no noise [9, 10] demonstrate that the CHC algorithm was unable to locate the global optimum for the F6 and F8 functions even once in the maximum trials allotted. However, Table 4.2 shows that the CHC algorithm is very effective in locating noisy solutions that approximate the optimal solution for the noisy F6 and F8 functions when using the noisy function evaluation performance metric. The CHC algorithm is also the most efficient and effective algorithm for solving the Schwefel function. Additionally, the CHC algorithm proved to be very robust and required no parameter tuning.

The average true fitness values in Table 4.2 were obtained for these experiments to evaluate how well the solutions produced by the noisy fitness function value metric approximate the known optimum solutions for the actual functions. These values, however, are subject to the range of solutions that satisfy the stopping criteria; the number of solutions satisfying these criteria depends on the fitness landscape of the functions.

4.5   Performance Comparisons Using True Fitness Values in Noisy Optimization Environments

The following tests were accomplished using the *true* fitness value of the offspring to determine when the optimal solution was located. The only feedback provided to the algorithms, however, was the noisy fitness function evaluation values.

Progress was measured with respect to each string's *true evaluation value*. As mentioned previously, using the noisy fitness function evaluation value to indicate that the algorithm has satisfied a stopping criterion allows a range of solutions to act as satisfactory solutions. This range is conditional on the noise component and the fitness landscape of the specific function. Measuring the performance of the algorithms by the *true* fitness value while providing the noisy feedback to the algorithms provides a different metric to determine how well the algorithm performs relative to the actual fitness function.

Gray coding was not used in these comparisons and the same algorithms are compared: the mutation driven stochastic hill-climber, delta coding, the CHC adaptive search algorithm, and the ESGA.

For the following performance comparisons the parameters for each algorithm were set to match those found in previous studies [10] and then tuned by judgment to give the best performance possible. The best possible performance for these tests was defined as: (1) the most number of runs locating the *true* optimal solution for the function while performing the fewest number of recombinations; or (2) when the problem is not solved consistently, the smallest average solution in the maximum number of trials allowed.

4.5.1   Empirical Results and Analysis

The performance results for each of the four algorithms tested on the functions in this test suite with Gaussian noise injected into the evaluation function are presented in Table 4.3. Percent solved indicates the percentage of the 30 independent runs that found the *true* optimal solution to the corresponding test function. Average trials indicates the number of recombinations performed by the algorithm for those runs that found the optimal solution. The value for average best indicates the average best solution found with respect to the true function evaluation (over 30 runs) after the maximum number of trials has been executed. Additionally, the lowest solution value is the lowest *true fitness value* found over the 30 runs for the corresponding test function.

The results presented in Table 4.3 indicate that the addition of Gaussian noise to these benchmark functions provides a very difficult test suite when the stopping criterion for the search is locating the *true* global optimum.

The stochastic hill-climber performs worse than all of the genetic algorithms tested here. The F4 results in Table 4.3 also indicate that the performance metric based on the *true* function evaluation value is much more difficult to satisfy than the stopping criteria using the noisy function evaluation value. In fact, the results in Table 4.1 would seem to indicate that the genetic algorithms tested here are consistently able to solve the F4 function. However, the results in Table 4.3 indicate that none of the algorithms tested here are able to locate the true optimal solution for the F4 function within the maximum trials allotted. This seems to imply that while the genetic algorithms are more tolerant of noise than the stochastic hill-climber tested here, they are not able to filter out all of the noise and consistently locate the optimal solution. This is consistent with the observation that the noise component of the evaluation function may dominate the feedback that the genetic algorithm receives as the population converges on strings with evaluations that are similar to the string representing the optimal solution.

These results also indicate that the CHC and delta coding genetic algorithms perform significantly better than the other algorithms tested here. CHC and delta coding are the only two algorithms able to solve the F7 function and both algorithms reach much better average solutions on the other functions of the test suite. This may suggest that these iterative algorithms (i.e., delta coding and CHC) are very effective in quickly locating solutions that are in the vicinity of the optimal solution. Nevertheless, their ability to locate the true optimal solution is diminished as the

Table 4.3 Algorithm performance comparisons on noisy functions using the true fitness function evaluation value for optimality

Algorithm	Problem (no. params)	F4 (30)	F6 (20)	F7 (10)	F8 (10)
	Total bits	240	200	100	100
	Maximum trials	100,000	500,000	200,000	500,000
Stochastic hill-climber	Percent solved	0%	0%	0%	0%
	Average best	0.883	19.028	−4,092.9	1.144
	σ best	0.340	5.148	93.315	0.161
	Lowest solution	0.346	10.720	−4,184.5	0.686
	Mutation rate	0.05	0.04	0.07	0.07
ESGA	Percent solved	0%	0%	0%	0%
	Average best	0.598	14.424	−4,181.2	1.029
	σ best	0.235	3.338	27.58	0.102
	Lowest solution	0.308	9.070	−4,189.6	0.761
	Population size	50	50	50	50
	Crossover rate	0.80	0.80	0.60	0.90
	Mutation rate	0.005	0.01	0.01	0.005
CHC	Percent solved	0%	0%	100%	0%
	Average trials			64,618	
	σ trials			17,122	
	Average best	0.068	1.101		0.585
	σ best	0.025	0.649		0.064
	Lowest solution	0.023	0.218		0.426
	Population size	50	50	50	50
	Cataclysmic mutation		35%	35%	35%
δ coding	Percent solved	0%	0%	90%	0%
	Average trials			45,706	
	σ trials			10,170	
	Average best	0.141	0.366	−4,182.0	0.465
	σ best	0.043	0.434	29.67	0.049
	Lowest solution	0.057	0.059	−4,189.8	0.391
	Population size	25	800	100	200
	Linear bias	2.00	1.90	1.90	2.00
	Mutation rate	0.01	0.01	0.02	0.005

Note: The *true* optimal solution for the F4, F6, and F8 functions is 0.0. For the F7 function the *true* optimal solution is −4,189.82764 (conditional upon the machine architecture).

variance in string evaluations becomes dominated by the noise component of the evaluation function. Neither of these two algorithms is clearly best for locating the *true* optimal solution in noisy optimization environments.

In previous comparisons where noise was not interjected, delta coding was the only algorithm tested that was able to consistently locate the optimal solution to the Rastrigin and Griewank functions without the use of Gray coding [10]. Additionally, delta coding performed significantly better than the other algorithms tested on the F4 function using the noisy function evaluation value metric. However, in these comparisons the delta coding algorithm was not able to locate the *true* optimal solution in any of the F4, F6, or F8 experiments in the trials allotted.

4.6   Discussion of Empirical Tests

The results presented here indicate that this test suite is resistant to hill-climbing. Noisy function evaluation significantly impacts the performance of the stochastic hill-climber tested here, while the noisy feedback has less severe effects on the performance of the population-based genetic search algorithms. During the early stages of genetic search noise seems to be effectively filtered so that the genetic algorithm can adequately sample the fitness landscape. However, as solutions that are competitive with the global optimum are discovered by the search, the noise component of the fitness function becomes an increasingly significant part of the individual's fitness. As the noise component begins to dominate the feedback to the algorithm it becomes more difficult to approximate the underlying fitness landscape. At this point in the search process strategies that more randomly sample the immediate neighborhood of solutions in the current population may have an advantage in locating the global optimum. This may explain why CHC performed so well in the experiments that used the true fitness value as the performance metric. One enhancement that might improve the performance of these algorithms with respect to locating the global optimum is to re-evaluate strings multiple times as the noisy fitness function evaluation values become comparable to the global optimum. This would help filter the noise and possibly allow the algorithms to better approximate the true fitness of the strings.

Of the algorithms tested here delta coding and the CHC adaptive search algorithm appear to be more effective in quickly locating quality solutions for the corresponding functions regardless of the performance criteria. Delta coding appears to perform slightly better than CHC when the noisy function evaluation value is used as a stopping criteria. However, the CHC adaptive search algorithm performed slightly better when the *true* fitness value was used to measure the performance of the algorithms.

The two sets of criteria used in these experiments as performance metrics demonstrate the dilemma of designing an adequate method for evaluating an algorithm's ability to search a noisy function. Requiring that the algorithm locate solutions that are within some σ limits of the optimal solution is unsatisfactory, as the number of solutions that meet this criterion is dependent on the fitness landscape of the underlying function.

Finally, it is important to note that these comparisons evaluate the performance of these algorithms for a small number of functions which have had only one type of noise artificially injected into the fitness evaluation function. Thus, any conclusions are speculative; this motivates the need to develop a larger test suite of functions that include the injection of different types of noise.

4.7 An Application: Geophysical Static Corrections

Images of subsurface strata obtained from seismic reflection surveys are often distorted due to *ambient noise* and the fact that different seismic signals travel through dissimilar materials near the earth's surface. The geophysical static corrections problem is defined as the search for the temporal offsets necessary to correct the signals obtained in the reflection surveys resulting in an accurate geologic strata image. This problem is very difficult because of the inherent signal noise, very large search spaces with many local optima, and highly nonlinear characteristics of the search space. Conventional methods for finding static corrections often result in unacceptable solutions when the level of ambient noise present is high or when severe surface heterogeneities exist. Therefore, global search methods that are not sensitive to *noise* have been explored.

Attempts to use genetic algorithms for solving the static corrections problem have previously proven to be computationally ineffectual [14, 19]. Even attempts to combine genetic search with local search have proven to be computationally expensive [16]. Nonetheless, these hybrid genetic algorithms have resulted in significantly improved solutions. These hybrid genetic algorithms use a *noisy* evaluation function that computes only a small percentage (i.e., 10%) of the cross-correlations between the signals in the reflection survey. This also introduces additional noise in the evaluation function.

4.7.1 Problem Description

Seismic reflection surveys consist of a series of signals generated by a detonation (shot) and recorded at a series of sensor sites. Figure 4.1a illustrates the reflection signals that a single sensor (□) would collect for a single shot (○) as well as the waveform corresponding to the signal. Each of the sensors along the sensor line would collect a similar set of reflection signals for this shot. A number of shots would be detonated over time, moving the shot site to the adjacent sensor site for each new signal collection (top of Figure 4.1b).

The signals collected at each sensor can be filtered and recombined so that all reflections occurring at common midpoints are grouped into corresponding synthesized traces. Figure 4.1b illustrates the reflection signals that would be grouped into a common midpoint (CMP) gather. These synthesized traces are then rotated 90° and grouped together to form a distorted image of the subsurface strata called a *stack*. Applying the true temporal corrections to each signal results in an accurate image of the subsurface strata.

The traces in a CMP gather are affected by two types of temporal phenomena. The first temporal effect is due to the difference in distance that the signals travel from various sensor/shot pairs. The temporal corrections for this phenomenon are collectively known as the *normal move out* (NMO) and can be approximated since the temporal differences between signals at a CMP are directly related to the distance traveled. The second type of temporal phenomenon is a result of the difference in materials at the earth's surface. Dense materials tend to increase the velocity of the signal, while porous and loosely packed materials tend to delay the signal. Since the materials beneath the detonation and sensor sites may vary, two independent "static" corrections are applied to each signal. The static corrections (one for each signal) make up the parameters that we wish to optimize.

4.7.2   Algorithm Descriptions

Our initial experiments, as well as the experiences of other researchers, indicated that some combination of steepest ascent and genetic search is needed to achieve adequate performance on this problem. We therefore compared the performance of four types of hybrid genetic algorithms on the geophysical static corrections problem. The descriptions of those algorithms follow.

4.7.2.1   The GENITOR Hybrid Algorithm

The GENITOR hybrid algorithm tested here uses a population of binary strings. When decoded, these strings represent the static corrections to the signals in the seismic reflection survey. The strings are initially seeded in the population using a small set of biased correction models (e.g., 5) that are calculated from *a priori* knowledge about the signal correlations. These models are then improved by iterative application of waveform steepest ascent. The improved models are then converted to binary representations to serve as templates for initializing the population. The binary strings in the population are then formed by repeatedly mutating 15% of the bits from the template solutions and placing the results in the population (i.e., cataclysmic mutation). None of the template strings are included in the population initially.

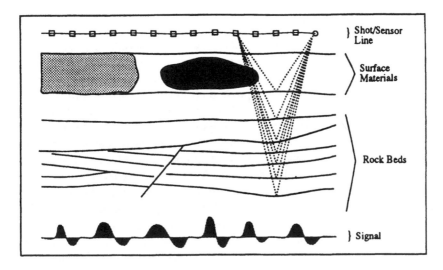

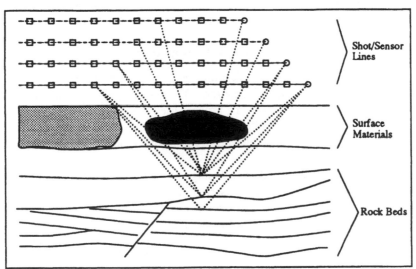

FIGURE 4.1
Sensor shot signals (top, a) and common midpoint gather (bottom, b).

A single NMO is used for the evaluation of all offspring. This NMO is computed by averaging the NMOs of the template solutions. The evaluation function returns the cross-correlation power of the stack. Genetic search proceeds without application of any local search methods until the population converges. Genetic search is executed using the GENITOR algorithm. When the population has converged, waveform steepest ascent is iteratively applied to the best solution in the population to improve the quality of the final solution.

4.7.2.2   The Staged Hybrid ESGA Model

The design of the *staged* hybrid ESGA resembles that of the hybrid GENITOR algorithm in that the two algorithms both intelligently seed their initial populations from solutions that have been improved by iterative application of waveform steepest ascent. They both also improve their final solutions by iterative application of the steepest ascent algorithm. However, the staged hybrid ESGA employs a floating point representation for the population individuals. This same representation was used by Stork and Kusuma [16], where each floating point number corresponds to a single static correction. Additionally, *all* of the individuals in the initial population were developed directly from *a priori* signal correlation knowledge and waveform steepest ascent was iteratively applied to all of the members of the population. This allows a unique NMO to be paired with each individual.

After the initial population has been formed, genetic search is allowed to continue uninterrupted for ten generations. After every tenth generation one iteration of waveform steepest ascent is applied to all individuals in the population. This pattern is repeated until the 40th generation when the waveform steepest ascent algorithm is applied iteratively to the entire population again.

4.7.2.3   The Staged Hybrid GENITOR Algorithm

The *staged* hybrid GENITOR algorithm employs binary string representations as used in the GENITOR hybrid algorithm. The initial population and initial NMO model are developed using the same process used in the GENITOR hybrid algorithm. Then genetic search is allowed to proceed. However, after some number of predefined trials have been executed (e.g., 10,000) the genetic search is temporarily halted.

The best solution at the end of each stage of the genetic search is converted to a floating point representation and inserted into the original set of models used to seed the population, displacing the worst model. At this point, one iteration of waveform steepest ascent is applied to the floating point models. These models are then used as templates to re-seed the population for the next stage of genetic search in the same manner as they were used in the initial seeding process. A single NMO is also developed for the evaluation of the offspring using the same averaging technique used initially. Genetic search is then restarted.

This cycle of genetic search followed by local search is repeated until some number of total trials have been executed. When the final stage of genetic search is completed, the best solution in the population is inserted into the floating point model set and waveform steepest ascent is iteratively applied to improve the final solution.

4.7.2.4   The Hybrid Delta Coding Algorithm

The hybrid version of delta coding (Figure 4.2) that we tested on the static corrections problem employed binary strings. The strings were initialized using the

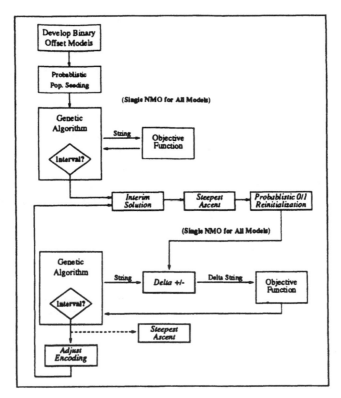

FIGURE 4.2
Hybrid delta coding algorithm.

same procedures described for the GENITOR hybrid and staged hybrid GENITOR algorithms. The initial NMO used for evaluating offspring was also developed using the same averaging technique as the other two algorithms. After the population is seeded, genetic search using the GENITOR algorithm is allowed to proceed for a predetermined number of trials. The normal method of testing population diversity using the Hamming distance metric was not used for testing on this problem. At the end of the genetic search iteration the best solution in the population is converted to a floating point representation and a single iteration of waveform steepest ascent is applied. This solution, along with the corresponding NMO, is saved as the interim solution. Each string in the following iteration is decoded and referenced with respect to this interim solution.

Population re-initialization is performed by producing binary strings where some percentage (i.e., 15%) of the bit positions in each string are initialized to ones. These positions are chosen randomly. One string consisting of all zeros is seeded into the population as well. This string represents the interim solution in the delta coded format. The other strings in the population, when decoded, represent static offsets close to the interim solution in numeric space.

After the population is re-initialized genetic search resumes. All offspring parameters are decoded as delta values ($\pm\delta$) which are added to the corresponding parameter in the interim solution. This allows the algorithm to explore a new mapping of the search space while employing the same objective function throughout the search. This iteration is allowed to proceed for the same number of predefined trials as the first stage of genetic search.

When the delta iteration is completed, the best solution is again converted to floating point format and waveform steepest ascent is applied. This results in a new interim solution. The population is then re-seeded using the method described for the second iteration of search except that the parameter representations are altered. The number of bits used to represent each parameter is reduced by one. This decreases the range of the δ values by half. When the population has been re-initialized, genetic search is resumed. This cycle is repeated until some number of trials have been executed. The final static corrections model is determined by iterative application of waveform steepest ascent to the best solution in the final population.

The hybrid delta coding algorithm provides the same separation of effort between the local and genetic search components as the other staged algorithms. It also saves computational effort by applying the waveform steepest ascent algorithm to a limited number of models. The delta coding algorithm also provides some additional advantages including a reduced population size (representing an additional time savings), reduced search space, and the exploration of multiple mappings of the search space over the course of the search.

4.7.3   Empirical Results and Analysis

Figure 4.3 shows typical runs of the basic GENITOR hybrid algorithm, a staged hybrid ESGA, a staged GENITOR hybrid algorithm, and a staged delta coding algorithm for the statics corrections problem. The evaluation value along the Y axis represents the power of the stack (which is being optimized) and the X axis indicates the time in CPU seconds as executed on a SPARC II workstation. The strategies shown here have been used for comparison due to their performance in other studies [12].

The staged hybrid GENITOR algorithm produces the best overall results for the static corrections problem. The objective function values produced by the staged hybrid GENITOR algorithm are approximately 7% better than those yielded by the staged hybrid ESGA and are produced approximately 3 hours faster. However, the delta coding algorithm produces results that are approximately 5% better than the staged hybrid ESGA but does so almost a full 5 hours faster.

Analysis of the performance curves that are associated with the staged search algorithms reveals that the sharp inclines indicating rapid improvement in the solutions being found are associated with the waveform steepest ascent stages of the search. The intervals indicating slow improvement are associated with the genetic stages of the search. This might lead one to believe that skipping the genetic search stages and iteratively executing local search would be advantageous. However, the genetic search stages actually support the local search by generating new points in different subpartitions of the search space, thus providing the local ascent algorithm with new points from which to climb. Furthermore, the best model associated with the waveform steepest ascent search may be a local optimum and, if so, this point would not be improved by additional local search.

4.8 Conclusions

The results presented here indicate that noisy function evaluation has more impact on the stochastic hill-climbing algorithm than on population-based genetic algorithms. Furthermore, dealing with noise during the early stages of genetic search seems to be relatively easy. But as the various genetic algorithms begin to generate solutions that were comparable to the global optimum, the noise component of the objective function becomes an increasingly significant factor during evaluation. At this point in the search process strategies that more randomly sample the immediate neighborhood of solutions in the current population may have an advantage in finding the global optimum. This may explain why CHC performed so well in our experiments.

We are thus left with a dilemma; the criteria we used may not be a good way to evaluate an algorithm's suitability for searching noisy functions; on the other hand, just requiring that an algorithm find some solution within 2 or 2.5 standard deviations of the evaluation of the global optimum also seems unsatisfactory. One alternative would be to use multiple evaluations (or more costly, but more actuate evaluation) as the search narrows its focus. This would provide a more direct means of canceling the noise at a point in the search when the variance over the set of solutions in the population may be dominated by the noise.

The delta coding algorithm proved to be slightly inferior to the staged hybrid GENITOR algorithm for the seismic "statics" problem in terms of the final results obtained, but delta coding found good solutions faster than the other algorithms we tested. We believe that delta coding was at a disadvantage in these experiments because it works with a single NMO rather than an average of several NMOs. Nevertheless, being able to generate good solutions quickly can also be advantageous in many application domains.

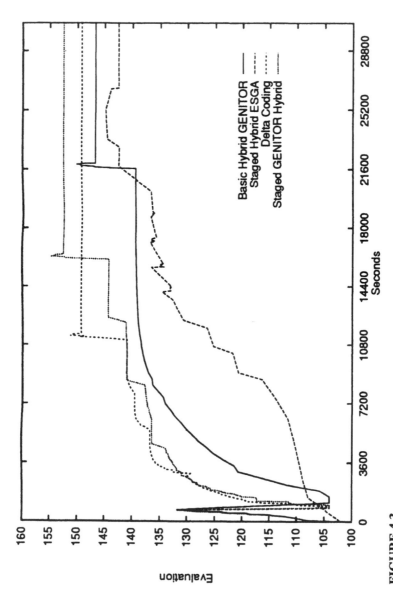

FIGURE 4.3
Timing performance comparisons.

Acknowledgments

This research was supported in part by a grant from the Colorado Advanced Software Institute and from NSF Grant IRI-9312748.

References

[1] Baker, J., Reducing bias and inefficiency in the selection algorithm, in *Genetic Algorithms and Their Applications: Proc. 2nd Int. Conf.*, Grefenstette, J., Ed., L. Elbraum Assoc., Hillsdale, NJ, 1987, 14.

[2] Booker, L., Improving search in genetic algorithms, in *Genetic Algorithms and Simulated Annealing*, Davis, L., Ed., Morgan Kaufmann, San Mateo, CA, 1987, 61.

[3] Caruana, R. and Schaffer, J., Representation and hidden bias: Gray vs. binary coding for genetic algorithms, in *Proc. 5th Int. Conf. on Machine Learning*, Morgan Kaufmann, San Mateo, CA, 1988.

[4] Davis, L., Bit-climbing, representational bias, and test suite design, in *Proc. 4th Int. Conf. on Genetic Algorithms*, Booker, L. and Belew, R., Eds., Morgan Kauffmann, San Mateo, CA, 1991, 18.

[5] Eshelman, L., The CHC adaptive search algorithm. How to have safe search when engaging in nontraditional generic recombination, in *Foundations of Genetic Algorithms*, Rawlins, G., Ed., Morgan Kaufmann, San Mateo, CA, 1991, 265.

[6] Fitzpatrick, J.M. and Grefenstette, J., Genetic algorithm in noisy environments, *Mach. Learning*, 3, 101, 1988.

[7] Goldberg, D., *Genetic Algorithms in Search, Optimization and Machine Learning*, Addison-Wesley, Reading, MA, 1989.

[8] Gordon, V. and Whitley, D., Serial and parallel genetic algorithms as function optimizers, in *Proc. 5th Int. Conf. on Genetic Algorithms*, Forrest, S., Ed., Morgan Kauffmann, San Mateo, CA, 1993, 177.

[9] Mathias, K.E. and Whitley, L.D., Changing representation during search: a comparative study of delta coding, *J. Evol. Comput.*, 2(3), 249–278, 1994.

[10] Mathias, K.E. and Whitley, L.D., Initial performance comparisons for the delta coding algorithm, in *Proc. IEEE Int. Conf. on Evolutionary Computation,* Schaffer, J.D., Ed., IEEE Press, Piscataway, NJ, 1994.

[11] Mathias, K.E. and Whitley, L.D., Transforming the search space with Gray coding, in *Proc. IEEE Int. Conf. on Evolutionary Computation,* Schaffer, J.D., Ed., IEEE Press, Piscataway, NJ, 1994.

[12] Mathias, K.E., Whitley, L.D., Stork, C., and Kusuma, T., Staged hybrid genetic search for seismic data imaging, in *Proc. IEEE Int. Conf. on Evolutionary Computation,* Schaffer, J.D., Ed., IEEE Press, Piscataway, NJ, 1994.

[13] Muhlenbein, H., Schomisch, M., and Born, J., The parallel genetic algorithm as a function optimizer, in *Proc. 4th Int. Conf. on Genetic Algorithms,* Booker, L. and Belew, R., Eds., Morgan Kauffmann, San Mateo, CA, 1991, 271.

[14] Scales, J., Smith, M., and Fischer, T., Global optimization methods of nonlinear inverse problems, in *Proc. Int. Conf. on Numerical Aspects of Wave Properties and Phenomenon,* France, 1990.

[15] Schaffer, J.D., Caruana, R.A., Eshelman, L.J., and Rajarshi Das, R., A study of control parameters affecting online performance of genetic algorithms for function optimization, in *Proc. 3rd Int. Conf. on Genetic Algorithms,* Schaffer, J.D., Ed., Morgan Kauffmann, San Mateo, CA, 1989, 51.

[16] Stork, C. and Kusuma, T., Hybrid genetic autostatics: new approach for large-amplitude statics with noisy data, in *Proc. SEG 62nd Annual Int. Meeting,* Society of Exploration Geophysicists, 1992, 1127.

[17] Whitley, L.D., The GENITOR algorithm and selective pressure: why rank based allocation of reproductive trials is best, in *Proc. 3rd Int. Conf. on Genetic Algorithms,* Schaffer, J.D., Ed., Morgan Kauffmann, San Mateo, CA, 1989, 116.

[18] Whitley, L.D., Mathias, K., and Fitzhorn, P., Delta coding: an iterative strategy for genetic algorithms, in *Proc. 4th Int. Conf. on Genetic Algorithms,* Booker, L. and Belew, R., Eds., Morgan Kauffman, San Mateo, CA, 1991, 77.

[19] Wilson, W. and Vasudevan, K., Application of the genetic algorithm to residual statics estimation, *Geophys. Res. Lett.,* 18(12), 2181, 1991.

5

Generalization of Heuristics Learned in Genetics-Based Learning

Benjamin W. Wah, Arthur Ieumwananonthachai, and Yong-Cheng Li

Abstract   In this chapter, we study methods for developing general heuristics in order to solve problems in knowledge-lean application domains with a large and possibly infinite problem space. Our approach is based on genetic learning that generates heuristics, tests each to a limited extent, and prunes unpromising ones from further consideration. We summarize possible sources of anomalies in performance evaluation of heuristics along with our methods for coping with them. Based on the heuristics learned, we propose and study methods for generalizing heuristics to unlearned problem domains. Our method uses a new statistical measure called probability of win, which assesses the performance of heuristics in a distribution-independent manner. To validate our approach, we show experimental results on generalizing heuristics learned for sequential circuit testing, VLSI cell placement and routing, and branch-and-bound search. We show that generalization can lead to new and robust heuristics that perform well across problem instances of different characteristics in an application domain.

5.1   Introduction

The design of problem solving algorithms for many applications generally relies on the expertise of designers and the amount of domain knowledge available. The design is difficult when there is little domain knowledge or when the environment under consideration is different from which the algorithm is applied. In designing efficient algorithms for these knowledge-lean applications, there are two important problems to be considered: (1) automated design of problem solving heuristics, and (2) systematic generalization of the heuristics learned. In this chapter, we focus on the second problem and present new methods for generalizing heuristics learned.

A *problem solver* can be optimal or heuristic. An optimal problem solver is a realization of an optimal algorithm that solves the problem for which it was designed optimally with respect to certain *objectives*. In contrast, a heuristic problem solver has components that were designed in an *ad hoc* fashion, leading to possibly suboptimal solutions when applied. When there is no optimal algorithm, the design of effective heuristics is crucial. Without ambiguity, we simply use "problem solvers" in this chapter to refer to "heuristic problem solvers".

Heuristics, in general terms, are "rules of thumb" or "common-sense knowledge" used in attempting the solution of a problem [14]. Newell et al. defined heuristics as "A process that may solve a given problem, but offers no guarantees of doing so" [12]. Pearl defined heuristics as "Strategies using readily accessible though loosely applicable information to control problem-solving processes in human being and machines" [14]. In this chapter, we define a *heuristic method (HM)* to mean a problem-solving procedure in a problem solver. Without loss of generality, a HM can be considered as a collection of interrelated *heuristic decision elements (HDE)* or *heuristics decision rules*. As illustrated in Figure 5.1, a heuristic problem solver takes a problem instance (or test case) and generates a possibly suboptimal solution.

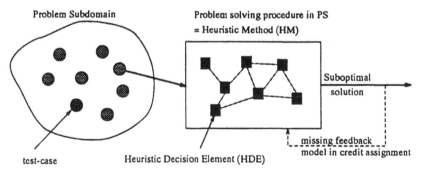

FIGURE 5.1
A heuristic method applied to a problem instance in a knowledge-lean application.

Heuristics are usually designed by human experts with strong expertise in the target application domain, or by automated systems using machine learning techniques. Both approaches focus on explaining the relation between heuristics and their performance, and on generating "good" heuristics based on observed information or explained relations. Regardless of the acquisition methods, heuristics in most cases are obtained by experimenting on a subset of test cases that cover a small portion of the problem space. For many applications, the problem space to be traversed is large, and there may be different regions that can best be solved by different heuristics. For this reason, we study in this chapter methods for generalizing heuristics learned for a knowledge-lean application to regions of its problem space not seen during learning.

Designing heuristics for knowledge-lean applications is usually based on a generate-and-test paradigm. In the following, we present issues involved in this process and our assumptions made in this chapter.

5.1.1 Generation of Heuristics

The way that heuristics are generated depends on domain knowledge available in the application environment. An application environment can be knowledge-rich or knowledge-lean with respect to the heuristics to be designed. In a knowledge-rich domain, a *world model* helps explain the relationship among decision, states, actions, and performance feedback generated by the learning system or measured in the environment. This model is important in identifying good heuristics that otherwise may be difficult to find. In contrast, such models do not exist in knowledge-lean domains. In this case, the heuristics generator cannot rely on performance feedback (or credit assignment as shown in Figure 5.1) to decide how new heuristics should be generated or how existing heuristics should be modified. Operators for composing new HMs for knowledge-lean domains are generally model-free, domain-independent, and syntactic in nature. A popular learning method using these operators is genetics-based learning.

Genetics-based learning is a generate-and-test paradigm that maintains a pool of competing HMs, tests them to a limited extent, creates new ones from those that perform well in the past, and prunes poor ones from the pool. It involves applying genetic algorithms [15] to machine learning problems and is most suitable for learning performance-related HMs. (Performance-related HMs aim to improve the performance of solutions for an application problem, whose constraints are trivially satisfied.) This is true because genetic operators do not rely on domain knowledge in generating heuristics and are not able to generate (correctness-related) heuristics for applications whose constraints are hard to satisfy. Examples of genetics-based methods for learning performance-related heuristics include population-based learning [23] and genetic programming [9].

As our focus in this chapter is on generalization, we assume that a set of "good" heuristics for some subsets of the problem space of an application has been obtained. Regardless of the learning method used, it is clear that learning can only focus on a small portion of the problem space, and heuristics learned this way are restricted to domains studied in learning.

5.1.2 Testing of Heuristics and Evaluating Their Performance

To evaluate HMs in a knowledge-lean application domain, they must be tested on a set of problem instances (or test cases). Performance in this case is generally assessed in a statistical sense, without requiring exhaustive tests of all test cases.

In this chapter, we are interested in two types of problem domains: (1) those with a large number of test cases and possibly an infinite number of deterministic HMs for solving them, and (2) those with a small number of test cases but the HMs concerned have a nondeterministic component, such as a random initialization point, that allows different results to be generated for each test case. In both types, the performance of a HM is nondeterministic, requiring multiple evaluations of the HM on different test cases (type 1) or multiple evaluations of the HM on the same test case (type 2). Consequently, we need to define valid statistical metrics for comparing two HMs without exhaustively testing all test cases using these HMs. This requires identifying subsets of test cases whose collective behavior on a HM can be evaluated statistically. Moreover, we must also deal with conditions in which performance values of HMs can have different ranges within a subset of test cases and have entirely different distributions across multiple subsets. We present in Section 5.2 issues on selecting appropriate methods for coping with anomalies[1] in performance evaluation of heuristics.

5.1.3 Generalization of Heuristics Learned to Unlearned Domains

Since the problem space is very large and learning can only cover a small subset, it is necessary to generalize HMs developed to test cases not studied during learning. Generalization is difficult when HMs do not perform consistently or have different ranges of performance across different test cases. This issue has been somewhat ignored in the literature on genetic algorithms and must be addressed in order to find general and efficient HMs for solving a wide range of problem instances of an application. This issue is the focus of this chapter and is presented in Section 5.3.

In short, we study in this chapter methods for generalizing performance-related heuristics for knowledge-lean applications. We assume that performance of a HM is represented by one or more statistical metrics and is based on multiple evaluations of test cases (noisy evaluations). The major issues we study include methods to cope with inconsistencies in performance evaluation of heuristics (Section 5.2.3), and generalization of learned heuristics to unlearned domains (Section 5.3). Experimental results on several real-world applications are shown in Section 5.4. Heuristics for these applications were learned using TEACHER, a genetics-based learning system we had developed earlier [23].

---

[1] An anomaly means that one HM is better than another using one evaluation method, but worse using another evaluation method.

5.2 Performance Evaluation and Anomalies

In problem solving, a problem solver applies a sequence of decisions defined in HDEs of a HM, one after another, until an input test case is solved. These decisions, initiated by the problem solver at decision points, change the state of the application environment that is evaluated by a number of user-defined performance *measurables*. The problem solver then uses the performance measured to make further decisions. A *solution* in this context is defined as a sequence of decisions made by the HM on an input test case to reach the final state.

The *performance* of a HM on a test case depends on the *quality* of the solution found by the HM for this test case as well as the *cost* (e.g., computation time) in finding the solution. Here, we define *quality* (respectively, *cost*) of a solution with respect to an input test case to be one or more measures of how good the final state is (respectively, how expensive it is to reach the final state) when the test case is solved, and be independent of intermediate states reached. Note that cost and quality are in turn defined as functions of measurables in the application environment. We call quality and cost examples of *performance measures* of the application.

In this section we discuss issues related to performance evaluation of HMs. We show that a HM can be found to be better or worse than another HM depending on the evaluation criterion. Such inconsistencies are called *anomalies* in this chapter and are attributed to different methods of evaluating performance and to different behavior of HMs under different conditions. We propose methods to cope with these anomalies. When such anomalies cannot be avoided, our system provides users with alternative HMs so that users can pick the best HM(s) to satisfy their requirements.

5.2.1 Example Applications

In problem solving, a problem solver applies some general domain-independent algorithms, that rely on information provided by domain-dependent heuristics to make decisions and to change the current state of the application environment. Table 5.1 shows examples of a few practical applications and their domain-dependent heuristics. We use the first three applications to test our generalization strategy in this chapter. Results on the generalizing HMs in process mapping [23], load balancing [11], and stereo vision [18] have been presented elsewhere.

The first application we have studied is based on two genetic-algorithm packages (CRIS [17] and GATEST [16]) for generating test patterns in VLSI circuit testing. Both packages use a domain-independent genetic algorithm [15] that continuously evolves test patterns by analyzing mutated vectors on their ability to identify (or cover) more faults in a circuit. There are a lot of possible domain-dependent HMs; however, in our experiments, we chose the domain-dependent HMs for CRIS as

Table 5.1 Examples of Knowledge-Lean Applications and Their Learnable Domain-Dependent Heuristics

Application	Objective(s)	Domain-dependent parametric-heuristics	Heuristic element(s)	Example(s) of element
Genetic search of the best VLSI test sequence [16, 17]	Maximize fault coverage	Controls used in the genetic algorithm: iteration, rejection ratio, sequence depth, control factor, frequency of usage	Numeric values, fitness function	(2, 3, 4, 3.2, 100), H(•)
Simulated annealing:Timber Wolf [19]	Minimize area of layout with fixed maximum number of layers	If ([acceptance ratio] > [threshold]), then reduce temperature to next lower value	Numerical threshold value, cost function, temperature function	0.9, C(•), T(•)
Branch-and-bound search for finding a minimum-cost tour in a graph	Minimize cost of tour, satisfy constraint on visiting each node exactly once	If a node has the smallest decomposition-function value among all active nodes, then expand this node	Symbolic formula	Lower bound + upper bound of node
Process mapping for placing a set of processes on a multicomputer [6]	Minimize overall completion time, minimize time to find such mappings	If (processor utilization/average utilization of all processors) > (threshold), then evict one process	Numeric threshold value	1.10
Load balancing in distributed systems [11]	Minimize completion time of an incoming job	If (average WL[•] > [threshold]), then migrate this process	Workload function WL, numeric threshold value	WL(•), 2.0
Stereo vision for depth perception [18]	Minimize error in range estimation	Marr and Poggio's iterative algorithm	(Channel width, low edge-detection threshold, high threshold)	(0.6, 2.0, 5.0)
Designing a blind equalizer	Minimize convergence time, accumulated errors and cost, maximize S/N ratio	Objective (error) function for gradient descent	Symbolic formula of the error function	E(•)

a set of seven numeric parameters and for GATEST [16], one of the four fitness functions for computing the performance of test patterns. Our results here extend the performance results we have found earlier for CRIS [23] and show the ability of our learning and generalization procedures to result in higher fault coverages than the original packages.

The second application we consider is TimberWolf (Version 6), [19, 20] a software package based on simulated annealing for placing and routing a set of VLSI circuit components. We illustrate that, by tuning a set of six numeric parameters in TimberWolf, we can reduce the area of the chip as well as the time needed to find the layout. Results on extending our results on TimberWolf's cost and temperature-control functions will be shown in a future publication.

The last application is a software package WISE [21] that implements a branch-and-bound search to find optimal solutions of three combinatorial optimization problems (vertex cover, asymmetric traveling salesman, and knapsack packing). In this case, the branch-and-bound search is domain-independent, and we chose as domain-dependent heuristics the decomposition HM in the search algorithm. The decomposition HM is used to pick an attribute to decompose a subproblem in a search tree into descendants. For instance, in a vertex-cover problem, the goal is to find the minimum number of nodes of a graph so that each edge is emanating from one of the covered nodes. In this case, a subproblem represents a set of nodes in the graph to cover partially the edges in the graph, and the decomposition HM picks the next node to be included in the covered set. The HM is represented as a symbolic formula of parameters that can be obtained in the search tree.

5.2.2 Problem Subspace and Subdomain

Within an application domain, different regions of the problem space may have different characteristics, each of which can best be solved by a unique HM [15]. Since learning is difficult when test cases are of different behavior and it is necessary to compare HMs quantitatively, we need to decompose the problem space into smaller partitions before learning begins. In the following we define a problem subspace and a problem subdomain.

A *problem subspace* is a user-defined partition of a problem space so that HMs for one subspace are learned independently of HMs in other subspaces. Such partitioning is generally guided by common-sense knowledge or by user experience in solving similar application problems. It requires knowing one or more attributes to classify test cases and is driven by a set of decision rules that identify the subspace to which a test case belongs. Such partitioning is important when test cases in an application have vastly different behavior. However, in some cases, it may not be possible to define the attributes needed for partitioning, or the number of attributes is too large. When these happen, nonparametric clustering methods, such as those based on neural networks, may have to be used. Another possibility is to always apply multiple HMs for each test case, resulting in a higher computational cost for a better solution.

For instance, consider solving a vertex-cover problem described in the last subsection. In designing a decomposition HM to decide which vertex to be included in the covered set, previous experience on other optimization problems indicates that HMs for densely connected graphs are generally different from HMs for sparsely connected ones. Consequently, the problem space of all graphs may be partitioned (in an *ad hoc* fashion) into a small number of subspaces based on graph connectivities and learned independently. As another example, in generating test patterns for VLSI circuits, previous experience shows that sequential circuits require tests that are different from those of combinatorial circuits. As a result, we can partition the problem space into two subspaces. However, we are not able to partition the subspace of sequential circuits into smaller subspaces as it is not clear which attributes should be used in this partitioning.

Given a subspace of test cases, we next define a subdomain. A *problem subdomain* in this chapter is a partitioning of a problem subspace into smaller partitions so that one or more HMs can be designed for each partition. The reason for this partitioning is to allow quantitative comparison of performance of HMs in a subdomain, which may be difficult across subdomains. Recall that in our definition, test cases belong to the same subspace when they can be solved by the same HM, but this does not imply that their performance can be compared directly. In comparing the performance of HMs, it is necessary to aggregate their performance values into a small number of performance metrics (such as average or maximum). Computing these aggregate metrics is not meaningful when performance values are of different ranges and distributions. Hence, a subdomain is a maximal partitioning of test cases in a subspace so that different HMs can be compared quantitatively based on their aggregate metrics. It is important to point out that performance values may need to be normalized before aggregated statistically. We discuss issues related to normalization in the next subsection.

In the same way that test cases are partitioned into subspaces, minimal domain knowledge should be used in knowledge-lean applications to partition test cases into subdomains. Further, the requirement of the statistical metric for quantitative performance comparison must be satisfied. For instance, performance values need to be independent and identically distributed (i.i.d.) when the average metric is computed.

Continuing with the example on the vertex-cover problem, a problem subdomain can be defined as random graphs with a certain degree of connectivity. The performance of a decomposition HM can be defined as the number of subproblems expanded by a branch-and-bound search normalized with respect to that of a baseline HM. Proper normalization (to be discussed in the next subsection) allows meaningful statistical comparison of HMs within a subdomain. A possible goal in learning for a subdomain is, therefore, to develop a rank ordered set of HMs, based on the average fraction of vertices to cover all the edges of random graphs generated in this subdomain. As another example, in generating test patterns for VLSI circuits, we may have to treat each circuit as an individual subdomain as we do not know the best set of attributes to classify circuits.

It should now be clear that, in general, there are infinitely many subdomains in an application problem, and learning can only be performed on a small number of them. Consequently, generalization of HMs learned for a small number of subdomains to other subdomains in the same subspace is critical. Informally, generalization entails finding a good HM for solving a randomly chosen test case in a subspace so that this HM has a high probability of performing better than other competing HMs for solving this test case. In some situations, multiple HMs may have to be identified and applied together at a higher cost to find a solution of higher quality. We discuss generalization in detail in Section 5.3.

To illustrate the concepts presented in this chapter, we show in Figure 5.2 the average symmetric speedups of three decomposition HMs used in a branch-and-bound search to solve vertex-cover problems. (The use of symmetric speedup is defined in Equation 5.2 later.) We treat all test cases to belong to one subspace, and graphs with the same average degree of connectivity are grouped into a subdomain. We applied genetics-based learning to find the five best HMs for each of three subdomains with connectivities 0.1, 0.35, and 0.6. Figure 5.2 shows the performance of the best HMs learned in each subdomain across all subdomains. We have also identified a single generalized HM among the 15 HMs learned using the method discussed in Section 5.3 and show its performance in Figure 5.2. We found that the generalized HM is not the top HM learned in each subdomain, indicating that the best HM in each subdomain may be too specialized to the subdomain. We have also found that generalization is possible in terms of average performance. We need to point out that the average performance should not be used as the sole indicator, as the variances of performance may differ from one subdomain to another.

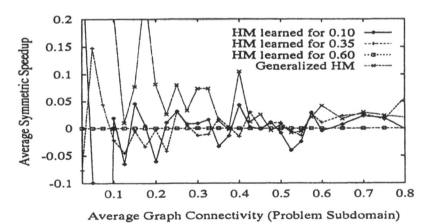

FIGURE 5.2
Average speedups (over 15 test cases) of three decomposition HMs for the vertex-cover problem, where subdomains are manually selected based on the connectivity of a graph. The HM learned for 0.6 connectivity is the same as the baseline HM. (Details of experiments are shown in Section 5.4.3)

5.2.3   Anomalies in Performance Evaluation

To design a good and general HM for an application, we must be able to compare HMs in terms of their performance across multiple problem subdomains within the same problem subspace. There are two steps in accomplishing this task. First, we must be able to compare HMs in terms of their performance within a single subdomain. Second, we must be able to compare the performance of HMs across multiple subdomains. Accomplishing the first step is necessary before we can deal with generalization in the second step. We have previously studied parts of the problem of evaluating performance of HMs within a single problem subdomain [22]. In this section, we summarize issues involved in the first step before presenting issues in the second step.

5.2.3.1   Anomalies within a Subdomain

Recall that HMs studied in this chapter have nondeterministic performance, implying the need to evaluate each HM multiple times in a subdomain. Further, performance may be made up of multiple measures (such as cost and quality) that are interrelated: higher quality is generally associated with higher cost.

To compare performance of different HMs, it is necessary to combine performance values before comparing them. This is, however, difficult, as the objectives of a HM as well as their trade-offs may be unknown with respect to its performance measures. A possible solution is to derive a single objective function of the performance measures with tunable parameters, and to find a combination of values of these parameters that lead to the HM with the best trade-off. Using this approach, we have observed the following difficulties before [22]:

- It is difficult to find a good combination of parameter values in the objective function so that HMs with the best quality-cost trade-offs can be found. We have seen similar difficulties in the goal attainment method [4].

- It is difficult to compare the performance of two HMs when they are evaluated on test cases of different sizes or behavior.

- Inconsistent conclusions (anomalies) about the performance of two HMs may be reached when the HMs are compared using either different user-defined objective functions or the same objective function with different parameters.

We have proposed before [23] three solutions to cope with these difficulties:

- Identify a *reference* or baseline HM upon which all other HMs are compared. A good choice of a reference method for a given application problem is the best existing HM.

- Normalize each raw performance measure of a new HM with respect to the same measure of the reference HM (evaluated on either the same set of test cases or test cases with the same distribution of performance) so that it is meaningful to compare two HMs based on their normalized measures.

- Compare two HMs based on individual normalized performance measures, not on a single parametric function of the measures.

In this section, we extend the anomalies found earlier [22] and classify all the anomalies into three classes. Note that anomalies happen because there is more than one dimension of performance variations.

a) Inconsistencies in Performance of HMs across Different Test Cases   When a HM is evaluated on a set of test cases, we must determine (1) the number of tests to be made and (2) the evaluation method (or metric) for aggregating performance values (such as mean, maximum, median, average rank). Inconsistent conclusions may be reached when one HM is better than another on one set of test cases, but worse on a different set of test cases.

For example, when performance is evaluated by the average metric, the ranking of HMs may change as more tests are performed. Similar observations have been found when HMs are evaluated by the average-rank metric.

In this chapter, we assume that all HMs are tested on the same set of test cases in a subdomain when evaluating generalizability, that performance values of tests in a subdomain are i.i.d., and that the average metric is used as the primary method for comparing HMs. In addition, we evaluate the actual distribution of performance values to avoid HMs that have good average behavior but have large spread in performance. When none of the HMs is a clear winner, our system will propose alternative HMs so that users can decide the appropriate HM(s) to use.

b) Multiple Objectives with Unknown Trade-Offs   The performance of a HM may be evaluated by multiple objectives (such as quality and cost). Of course, we would like to find HMs with improved quality and reduced cost. However, this may not always be possible, as improved quality is often associated with increased cost. The problem, generally known as a *multiobjective optimization problem* [4], involves trade-offs among all objectives of a design.

We have found before that anomalies may happen when HMs are optimized with respect to a *single* parametric function of the objectives of an application [7, 22]. The performance of a HM (as defined by the parametric objective function) may change drastically when minor changes are made on the parameters of the objective function. In fact, it is possible to show that one HM is better than another by finding a new parametric objective function of performance measures.

To avoid inconsistencies due to multiple objectives, we must evaluate HMs based on individual performance measures and not combine multiple measures into a single parametric function [22]. During learning, the learning system should constrain all but one measure and optimize the single, unconstrained measure.

A HM is pruned from further testing when one of its performance constraints is violated [23]. If a good HM satisfying the constraints can be found, then the constraints are further refined and learning is repeated. The difficulty with this approach is on setting constraints. We have studied the case when there are two performance measures [23]. However, the general case when there are more than two performance measures is still open at this time.

The applications we have studied in this chapter have only one performance measure to be optimized; hence, we do not need to deal with the issue of multiple objectives. (For GATEST, CRIS, and TimberWolf, the performance measure to be optimized is the quality of the result; for branch-and-bound search, the measure to be optimized is the cost of finding the optimal solution.)

c) Inconsistencies in Normalization   Normalization involves choosing a baseline HM and computing relative performance values of a new HM on a set of tests in a subdomain by the corresponding performance values of the baseline HM. This is necessary when performance is assessed by evaluating multiple test cases (type 1 as discussed in Section 5.1) and is not needed when nondeterminism in performance is due to randomness in the problem solver (type 2 as discussed in Section 5.1). In the former case, performance values from different tests may be of different ranges and distributions, and normalization establishes a reference point in performance comparison. In the latter, raw performance values within the subdomain are from one test case and presumably have the same distribution.

Normalization may lead to inconsistent conclusions about the performance of HMs when multiple normalization methods are combined or when normalization compresses/decompresses the range of performance values. These anomalies are explained as follows.

Inconsistencies in evaluation may occur when using multiple normalization methods as compared to using one normalization method. To illustrate these inconsistencies, consider with two HMs that have completion times $\{1475, 1665, 1381\}$ for HM_1 and $\{1269, 1513, 1988\}$ for HM_2 (Example A). Using HM_1 as the baseline for normalization, we can compute the average normalized speedup of HM_2 using one of the following methods:

$$\bar{q}_2^n = \sum_{j=1}^{3} \frac{t_{1,j}}{t_{2,j}} = 0.986; \qquad \bar{Q}_2^n = \sum_{j=1}^{3} \frac{t_{1,j} - 1200}{t_{2,j} - 1200} = 1.900 \qquad (5.1)$$

where $t_{i,j}$ is the completion time of HM i on test case j. Since the average normalized speedup of HM_1 is one, HM_2 is found to be worse using the first methods and better using the second.

Inconsistencies may also occur when normalization overemphasizes or deemphasizes performance changes. For instance, the conventional speedup measure is biased against slowdown (as slowdowns are in the range between 0 and 1, whereas speedups are in the range between 1 and infinity). As another example (Example B), suppose the speedups of a HM on two test cases are 10 and 0.1. Then the average speedup is 5.05, and the average slowdown is also 5.05, where the average slowdown is defined as the average of the reciprocals of speedup. In this case, both the average speedups and average slowdowns are greater than one.

In general, when normalizing performance values, it is important to recognize that the ordering of HMs may change when using a different normalization method, and that the spread of (normalized) performance values may vary across subdomains in an application. Here, we propose three methods to cope with anomalies in normalization. First, we should use only one normalization method consistently throughout learning and evaluation, thereby preserving the ordering of HMs throughout the process. Second, we need to evaluate the spread of normalized performance values to detect bias. This can be done by detecting outlyers and by examining higher-order moments of the performance values. Third, to avoid placing unequal emphasis on normalized values, we need a normalization method that gives equal emphasis to improvement as well as degradation. To simplify understanding, we describe this symmetric normalization method using the speedup measure. We define *symmetric speedup* as

$$Speedup_{symmetric} = \begin{cases} Speedup - 1 & \text{if } Speedup \geq 1 \\ 1 - \frac{1}{Speedup} & \text{if } 1 > Speedup \geq 0 \end{cases} \quad (5.2)$$

where speedup is the ratio of the time of the original HM with respect to the time of the new HM. Note that slowdown is the reciprocal of speedup, and that symmetric speedup is computed for each pair of performance values. Equation 5.2 dictates that speedups and slowdowns carry the same weight: speedups are in the range from zero to infinity, and slowdowns are in the range from zero to negative infinity.

In a similar way, we can define *symmetric slowdown* as

$$Slowdown_{symmetric} = \begin{cases} Slowdown - 1 & \text{if } Slowdown \geq 1 \\ 1 - \frac{1}{Slowdown} & \text{if } 1 > Slowdown \geq 0 \end{cases}$$

$$= -Speedup_{symmetric} \quad (5.3)$$

It is also easy to prove that $Speedup_{symmetric} = -Slowdown_{symmetric}$, thereby eliminating the anomalous condition in which average speedup and average slowdown are both greater than one or both less than one.

In Example A discussed earlier, the average symmetric speedup is -0.059, which shows that HM_2 is worse than HM_1. (The average symmetric slowdown is 0.059.) In Example B, both the average symmetric speedup and average symmetric slowdown are zero, hence avoiding the anomaly where the average speedup and average slowdown are both greater than one.

To illustrate the difference between speedups and symmetric speedups, we show in Figure 5.3 the distributions of speedups as well as symmetric speedups of a HM to solve the vertex-cover problem.

Table 5.2   Inconsistent Behavior of HMs in Different Subdomains

| Circuit ID | Heuristic method | | | | Fault coverages (%) | |
	ID	61801	98052	15213	Maximum	Average
		Random seeds used in HM				
s444	101	60.3	13.9	11.2	60.3	28.5
	535	81.9	86.3	86.3	86.3	84.8
s1196	101	93.2	94.4	94.9	94.9	94.2
	535	93.2	92.5	93.6	93.6	93.1

5.2.3.2 Anomalies across Subdomains

We now discuss the difficulty in comparing performance of HMs across multiple subdomains. This comparison is difficult when there is a wide discrepancy in performance across subdomains.

To illustrate this point, consider the three HMs shown in Figure 5.2. These HMs behave differently in different subdomains: not only can the range and distribution of performance values be different, but a good HM in one subdomain may not perform well in another. As another example, consider the HMs learned for CRIS (described in Section 5.2) [17]. A HM in this case is a vector of seven parameters and a random seed; by varying the random seed, we get different performance of the HM. Since it is difficult to characterize circuits with respect to their HMs, we treat the subspace to be all possible circuits, and each circuit as a subdomain for learning. The goal of generalization is to find one single set of parameters (HM) that gives high fault coverages for all circuits. In Table 5.2, we show two HMs and their fault coverages. With respect to circuit s444, HM_{101} has worse fault coverages and a wider distribution of coverage values than HM_{535}. On the other hand, HM_{101} performs better for circuit s1196.

The major difficulty in handling multiple problem subdomains is that performance values from different subdomains cannot be aggregated statistically. For instance, it is not meaningful to find the average of two different distributions. Scaling and normalization of performance values are possible ways to match the difference in distributions, but will lead to new inconsistencies for reasons discussed in Section 5.2.3.1.c. Another way is to rank HMs by their performance values across different subdomains, and use the average ranks of HMs for comparing HMs. This does not work well because it does not account for actual differences

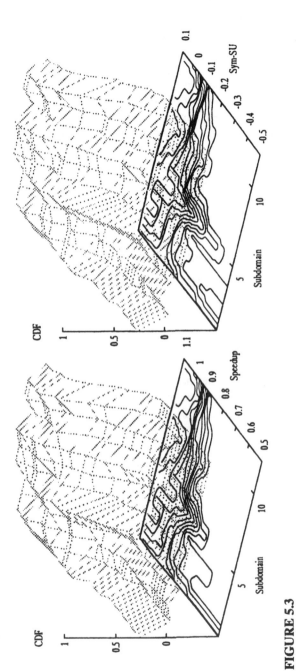

FIGURE 5.3
Distribution of performance values of one HM on 15 test cases based on speedup and symmetric speedup for solving the vertex-cover problem.

in performance values, and two HMs with very close or very different performance may differ only by one in their ranks. Further, the maximum rank of HMs depends on the number of HMs evaluated, thereby biasing the average ranks of individual HMs.

To address this problem, we propose in Section 5.3 a new metric called *probability of win*. Informally, the probability of win is a range-independent metric that evaluates the probability that the *true mean performance* of a HM in one subdomain is better than the true mean performance of another randomly selected HM in the same subdomain. The advantage of using probabilities of win is that they are in the range between zero and one, independent of the number of HMs evaluated and the range and distribution of performance values.

5.3   Generalization of Heuristic Methods Learned

The ability to generalize a learned HM to test cases in new problem subdomains is one of the key reasons for learning. Generalization is important because we perform learning on a very small number of subdomains, and there may be infinitely many subdomains in an application. Further, it is desirable to have one or very few HMs to be used in an application rather than a new HM for each problem instance.

The goal of generalization is somewhat vague: we like to find one or more HMs that perform well most of the time across multiple subdomains as compared to the baseline HM (if it exists). To achieve this goal, four issues are apparent here.

1.   How to compare the performance of HMs within a subdomain in a range-independent and distribution-independent fashion

2.   How to define the notion that one HM performs well across multiple subdomains

3.   How to find the condition(s) under which a specific HM should be applied

4.   What the trade-offs are between cost and quality in generalization

5.3.1   Probability of Win within a Subdomain

There are many ways to address issue (1) raised in this section, and the solutions to the remaining problems depend on this solution. As discussed at the end of the last section, ranking, scaling, and normalization do not work well. In this section, we propose a metric called probability of win to select good HMs across multiple subdomains.

P_{win}, the *probability-of-win* of HM h_i within a subdomain, is defined as the probability that the true mean of h_i (with respect to one performance measure) is better than the true mean of HM h_j randomly selected from the pool. When h_i is applied on test cases in subdomain d_m, we have

$$P_{win}(h_i, d_m) = \frac{1}{|s|-1} \sum_{j \neq i} P\left[\mu_i^m > \mu_j^m \,\middle|\, \hat{\mu}_i^m, \hat{\sigma}_i^m, n_i^m, \hat{\mu}_j^m, \hat{\sigma}_j^m, n_j^m\right] \quad (5.4)$$

where $|s|$ is the number of HMs under consideration, d_m is a subdomain, and n_i^m, $\hat{\sigma}_i^m$, $\hat{\mu}_i^m$, and μ_i^m are, respectively, the number of tests, sample standard deviation, sample mean, and true mean of h_i in d_m.

Since we are using the average performance metric, it is a good approximation to use the normal distribution as a distribution of the sample average. The probability that h_i is better than h_j in d_m can now be computed as follows.

$$P\left(\mu_i^m > \mu_j^m \,\middle|\, \hat{\mu}_i^m, \hat{\sigma}_i^m, n_i^m, \hat{\mu}_j^m, \hat{\sigma}_j^m, n_j^m\right) = 1 - \Phi\left[\frac{\hat{\mu}_i^m - \hat{\mu}_j^m}{\sqrt{\hat{\sigma}_i^{m2}/n_i^m + \hat{\sigma}_j^{m2}/n_j^m}}\right]$$

$$(5.5)$$

where $\Phi(x)$ is the cumulative distribution function for the $N(0, 1)$ distribution.

Table 5.3 Probabilities of Win of Four HMs

h_1	$\hat{\mu}_i$	$\hat{\sigma}_i$	n_i	$P_{win}(h_i)$
1	43.2	13.5	10	0.4787
2	46.2	6.4	12	0.7976
3	44.9	2.5	10	0.6006
4	33.6	25.9	8	0.1231

To illustrate the concept, we show in Table 5.3 probabilities of win of four HMs tested to various degrees. Note that the probability of win is not directly related to sample mean, but rather depends on sample mean, sample variance, and number of tests performed.

P_{win} defined in Equation 5.4 is range-independent and distribution-independent because all performance values are transformed into probabilities between 0 and 1. It assumes that all HMs are i.i.d. and takes into account uncertainty in their sample averages (by using the variance values); hence, it is better than simple scaling which only compresses all performance averages into a range between 0 and 1.

5.3.2   Probability of Win across Subdomains

The use of probability of win in a subdomain leads to two ways to solve issue
(2) posted earlier in this section, namely, how to define the notion that one HM
performs better than another HM across multiple subdomains. We present below
two ways to define this notion.

First, we assume that when HM h is applied over multiple subdomains in partition
Π_p of subdomains, all subdomain are equally likely. Therefore, we compute the
probability of win of h over subdomains in Π_p as the average probability of win
of h over all subdomain in Π_p.

$$P_{win}\left(h, \Pi_p\right) = \frac{\sum_{d \in \Pi_p} P_{win}\left(h, d\right)}{\left|\Pi_p\right|} \tag{5.6}$$

where Π_p is the pth partition of subdomains in the subspace. The HM picked is
the one that maximizes Equation 5.6. HMs picked this way usually win with a
high probability across most of the subdomains in Π_p, but occasionally may not
perform well in a few subdomains.

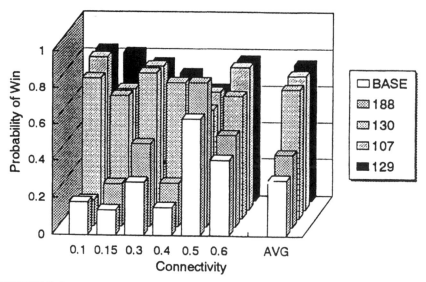

FIGURE 5.4
Histogram showing probabilities of win of four HMs generalized across six
subdomains and those of the baseline HM. (HM_{129} **is picked if Equation 5.6**
is used as the selection criterion; HM_{107} **is selected if Equation 5.7 is used as**
the criterion.)

Second, we consider the problem of finding a good HM across multiple subdomains in Π_p as a multiobjective optimization problem itself. As is indicated in
Section 5.2.3, evaluating HMs based on a combined objective function (such as the

average probability of win in Equation 5.6) may lead to inconsistent conclusions. To alleviate such inconsistencies, we should treat each subdomain independently and find a common HM across all subdomains in Π_p satisfying some common constraints. For example, let δ be the allowable deviation of the probability of win of any chosen HM from q_{win}^m, the maximum P_{win} in subdomain m. Generalization, therefore, amounts to finding h that satisfies the following constraints for every subdomain $m \in \Pi_p$.

$$P_{win}(h, m) \geq \left(q_{win}^m - \delta\right) \quad \text{for every } m \in \Pi_p \qquad (5.7)$$

In this formulation, δ may need to be refined if there are too many or too few HMs satisfying the constraints.

To illustrate the generalization procedure, consider the vertex-cover problem discussed in Section 5.2. Assume that learning had been performed on six subdomains (using graphs with connectivities 0.1, 0.15, 0.3, 0.4, 0.5, and 0.6, respectively), and that the five best decomposition HMs from each subdomain were reported. After full evaluation of the 30 HMs across all six subdomains, we computed the probability of win of each HM in every subdomain. Figure 5.4 shows the probabilities of win of several of these HMs. If we generalize HMs based on Equation 5.6, then HM_{129} will be picked since it has the highest average P_{win}. In contrast, if we generalize using Equation 5.7, then HM_{107} will be picked because it has the smallest deviation from the maximum P_{win} in each subdomain. Note that either HM is a reasonable generalized HM to be applied across all subdomains. To decide on the single generalized HM to use, further evaluations on the spread of performance values would be necessary (see Section 5.4).

Using probabilities of win to assess HMs across subdomains, we can now address issues 3 and 4 raised earlier in this section, which deal with the selection of multiple HMs. There are two ways that multiple HMs can be used: (1) using each HM in a nonoverlapping subset of subdomains in the subspace (issue 3), and (2) applying multiple HMs in solving any test case in the subspace (issue 4).

The issue on finding condition(s) under which a specific HM should be applied is a difficult open problem at this time. Solving this problem amounts to designing decision rules to partition the subspace of test cases into a finite number of partitions, each of which can be solved by a unique HM. This is possible in some applications that are characterized by a small number of well-defined attributes. For instance, in the vertex-cover problem discussed in Sections 5.2 and 5.4, the graph connectivity is a unique attribute that allows us to decompose the space of all random graphs into partitions. This is, however, difficult for applications that do not have a well-defined set of characteristic attributes. As a result, we cannot partition the subspace and assign a good HM to each. This happens in the CRIS test-pattern generation system [17] discussed in Sections 5.2 and 5.4.1 which does not have a set of characteristic attributes to classify circuits (e.g., number of gates, length of the longest path).

Finally, issue (4) raised earlier in this section is on trade-offs between cost and quality in generalization. Since it may be difficult in some cases to partition a subspace or to find a single HM that performs well across all test cases in the subspace, we can pick multiple HMs, each of which works well for some subdomains in the subspace, and apply all of them when a new test case is encountered. This is practical only when the added cost of applying multiple HMs is compensated by the improved quality of the solutions. In this approach, the cost reflects the total computational cost of applying *all* the chosen HMs to solve a given test case.

The problem of selecting a set of HMs for a subspace amounts to picking *multiple* HMs and assigning each to a subdomain in the subspace. Assuming that $|H|$ such HMs are to be found, we need to decompose the subspace into $|H|$ partitions of subdomains and assign one HM to all subdomains in each partition. The overall probability of win over the subspace is computed in a similar way as in Equation 5.6. In mathematical form, let Ω be the set of all HMs tested in the subspace and Π be the set of all subdomains in this subspace; we are interested to find $H \subset \Omega$ such that $|H|$ is constant and the average P_{win} is maximized. That is,

$$P_{win}^{max}(\Omega, \Pi) = \frac{\displaystyle\max_{\substack{H \subset \Omega \\ |H|=\text{constant}}} \sum_{d \in \Pi} \max_{h \in H} P_{win}(h, d)}{|\Pi|} \qquad (5.8)$$

where $|\Pi|$ is the number of subdomains in subspace Π.

One way to find H in Equation 5.8 is to enumerate over all possible ways of decomposing Π and assign the best HM to each partition. The problem is equivalent to the minimum-cover problem [3]: given a set Π of subdomains and a set Ω of HMs (each of which covers one or more subdomains), find the minimum subset H of Ω so that each element of Π can be covered by one HM in H. The problem is NP complete, in general, and is solvable in polynomial time only when $|H|$ is two.

Fortunately, by applying Equation 5.7, we can make the number of HMs arbitrarily small by choosing a proper δ. In this case, finding a fixed set of HMs that can best cover all subdomains in the subspace can be obtained by enumeration. Experimental results on such cost-quality trade-offs are presented in Section 5.4.1.

5.3.3 Generalization Procedure

The procedure to generalize HMs learned for subdomains in a problem subspace is summarized as follows:

1.   Using the collective set of HMs obtained in the subdomains learned, find the probability of win (using Equation 5.4) of each HM in every subdomain learned or to be generalized.

2. Apply Equation 5.8 to select the necessary number of HMs for evaluating test cases in the subspace. Equation 5.7 can be used to restrict the set of HMs considered in the selection process.

5.4 Experimental Results

To illustrate the generalization procedure described in Section 5.3.3, we present in this section results on generalization for the three applications discussed in Section 5.2.1.

5.4.1 Heuristics for Sequential Circuit Testing

The first application is based on CRIS [17] and GATEST [16], two genetic-algorithm software packages for generating patterns to test sequential VLSI circuits. In our experiments, we used sequential circuits from the ISCAS89 benchmarks [1] plus several other larger circuits. Since these circuits are from different applications, it is difficult to classify them by some common features. Consequently, we treat each circuit as an individual subdomain. As we like to find one common HM for all circuits, we assume that all circuits are from one subspace.

5.4.1.1 CRIS

CRIS [17] is based on continuous mutations of a given input test sequence and on analyzing the mutated vectors for selecting a test set. Hierarchical simulation techniques are used in the system to reduce memory requirement, thereby allowing test generations for large VLSI circuits. The package has been applied successfully to generate test patterns with high fault coverages for large combinatorial and sequential circuits.

CRIS in our experiments is treated as a problem solver in a black box, as we have minimal knowledge in its design. A HM targeted for improvement is a set of eight parameters used in CRIS (see Table 5.4). Note that parameter P_8 is a random seed, implying that CRIS can be run multiple times using different random seeds in order to obtain better fault coverages. (In our experiments, we used a fixed sequence of ten random seeds.)

A major problem in using the original CRIS is that it is hard to find proper values for the seven parameters (excluding the random seed) for a particular circuit. The designer of CRIS manually tuned these parameters for each circuit, resulting in HMs that are hard to generalize. This was done because the designer wanted to obtain the highest possible fault coverage for each circuit, and computation cost was only a secondary consideration. Note that the times for manual tuning were exceedingly high and were not reported in the Reference [17].

Table 5.4 Parameters in CRIS Treated as a HM in Learning and in
Generalization

Parameter	Type	Range	Step	Definition	Learned value
P_1	Integer	1–10	1	Related to the number of stages in a flip-flop	1
P_2	Integer	1–40	1	Related to the sensitivity of changes of state of a flip-flop (number of times a flip-flop changes its state in a test sequence)	12
P_3	Integer	1–40	1	Selection criterion—related to the survival rate of a candidate test sequence in the next generation	38
P_4	Float	0.1–10.0	0.1	Related to the number of test vectors concatenated to form a new test sequence	7.06
P_5	Integer	50–800	10	Related to the number of useless trials before quitting	623
P_6	Integer	1–20	1	Number of generations	1
P_7	Float	0.1–1.0	0.1	How genes are spliced in the genetic algorithm	0.1
P_8	Integer	Any	1	Seed for the random number generator	—

Note: The type, range, and step of each parameter were recommended to us by the designer of CRIS. The default parameters were not given to us as they are circuit-dependent.

Our goal is to develop one common HM that can be applied across all the benchmark circuits and that has similar or better fault coverages as compared to those of the original CRIS. The advantage of having one HM is that it can be applied to new circuits without further manual tuning.

5.4.1.2 GATEST

GATEST [16] is another test-pattern generation package based on genetic algorithms. It augments existing techniques in order to reduce execution times and to improve fault coverages. The genetic-algorithm component evolves candidate test vectors and sequences, using a fault simulator to compute the fitness of each candidate test. To improve performance, the designers manually tuned various genetic-algorithm parameters in the package (including alphabet size, fitness function, generation gap, population size, and mutation rate) as well as selection and crossover schemes. High fault coverages were obtained for most of the ISCAS89 sequential benchmark circuits [1], and execution times were significantly lower in most cases than those obtained by HITEC [13], a deterministic test-pattern generator.

The entire GA process was divided into four phases, each with its own fitness function that had been tuned by the designers manually. The designers also told us that phase 2 has the largest impact on performance and recommended that we improved it first. As a result, we treat GATEST as our problem solver, and the fitness function (a symbolic formula) in phase 2 as our HM. The original form of this fitness function is

$$fitness_2 = \#\_faults\_detected + \frac{\#\_faults\_propagated\_to\_flip\_flops}{(\#\_faults)\,(\#\_flip\_flops)} \qquad (5.9)$$

In learning a new fitness function, we have used the following variables as possible arguments of the function: *#_faults_detected, #_faults_propagated_to _flip_flops, #_faults, #_flip flops, #_circuit nodes,* and *sequence_length.* The operators allowed to compose new fitness functions include $\{+, -, *, /\}$.

5.4.1.3   Experimental Results

In our experiments, we chose five circuits as our learning subdomains. In each of these subdomains, we used TEACHER [23] to test CRIS 1000 times with different HMs, each represented as the first seven parameters in Table 5.4. At the end of learning, we picked the top 20 HMs and evaluated them fully by initializing CRIS by ten different random seeds (P_8 in Table 5.4). We then selected the top five HMs from each subdomain, resulting in a total of 25 HMs supplied to the generalization phase. We evaluated the 25 HMs fully (each with ten random seeds) on the five subdomains used in learning and five new subdomains. We then selected one generalized HM to be used across all the ten circuits (based on Equation 5.8). The elements of the generalized HM found are shown in Table 5.4.

For GATEST, we applied learning to find good HMs for six circuits (s298, s386, s526, s820, s1196, and s1488 in the ISCAS89 benchmark). We then generalized the best 30 HMs (5 from each subdomain) by first evaluating them fully (each with ten random seeds) on the six subdomains and by selecting one generalized HM for all circuits (using Equation 5.8). The final fitness function we got after generalization is

$$fitness_2 = 2 \times \#\_faults\_propagated\_to\_flip\_flops - \#\_faults\_detected. \qquad (5.10)$$

Table 5.5 shows the results after generalization for CRIS and GATEST. For each circuit, we present the average and maximum fault coverages (over ten random seeds) and the corresponding computational costs. These fault coverages are compared against the published fault coverages of CRIS [17] and GATEST [16] as well as those of HITEC [13]. Note that the maximum fault coverages reported in

Table 5.5 Performance of HMs in Terms of Computational Cost and Fault Coverage for CRIS and GATEST

Circuit ID	Total faults	Fault coverage				Cost		CRIS generalized HM			GATEST generalized HM		
		HITEC	CRIS	Avg GATEST	Max GATEST	HITEC	Avg GATEST	Avg FC	Max FC	Avg cost	Avg FC	Max FC	Avg cost
*s298	308	86.0	82.1	85.9	86.0	15,984.0	128.6	84.7	86.4	10.9	85.9	86.0	126.4
s344	342	95.9	93.7	96.2	96.2	4,788.0	134.8	96.1	96.2	21.8	96.2	96.2	133.3
s349	350	95.7	—	95.7	95.7	3,132.0	136.9	95.6	95.7	21.9	95.7	95.7	128.3
+s382	399	90.9	68.6	87.0	87.5	43,200.0	203.3	72.4	87.0	7.2	87.0	87.5	208.9
s386	384	81.7	76.0	76.9	77.9	61.8	67.6	77.5	78.9	3.5	78.6	79.3	78.6
*s400	426	89.9	84.7	85.7	86.6	43,560.0	229.3	71.2	85.7	8.4	85.7	86.6	215.1
s444	474	87.3	83.7	85.6	86.3	57,960.0	259.4	79.8	85.4	9.3	85.6	86.3	233.8
*s526	555	65.7	77.1	75.1	76.4	168,480.0	333.4	70.0	77.1	10.0	75.5	77.3	302.7
s641	467	86.5	85.2	86.5	86.5	1,080.0	181.2	85.0	86.1	19.5	86.5	86.5	195.0
+s713	581	81.9	81.7	81.9	81.9	91.2	219.9	81.3	81.9	23.0	81.9	81.9	256.5
s820	850	95.6	53.1	60.8	68.0	5,796.0	266.4	44.7	46.7	51.3	69.3	80.9	225.4
*s832	870	93.9	42.5	61.9	66.8	6,336.0	265.8	44.1	45.6	44.6	66.9	72.8	251.0
s1196	1,242	99.7	95.0	99.2	99.5	91.8	292.1	92.0	94.1	20.0	99.2	99.4	421.7
*s1238	1,355	94.6	90.7	94.0	94.4	132.0	380.5	88.2	89.2	23.0	94.0	94.2	585.2
s1488	1,486	97.0	91.2	93.7	96.0	12,960.0	512.3	94.1	95.2	85.6	94.3	96.5	553.4
+s1494	1,506	96.4	90.1	94.0	95.8	6,876.0	510.4	93.2	94.1	85.5	93.6	95.6	584.3
s1423	1,515	40.0	77.0	81.0	86.3	—	3,673.9	82.0	88.3	210.4	81.3	87.3	4,325.7
+s5378	4,603	70.3	65.8	69.5	70.1	—	9,973.3	65.3	69.9	501.8	69.6	71.9	8,875.7
s35932	39,094	89.3	88.2	89.5	89.7	13,680.0	184,316.0	77.9	78.4	4,265.7	89.4	89.7	184,417.0
am2910	2,573	85.0	83.0	—	—	—	—	83.7	85.2	307.6	—	—	—
+div16	2,147	72.0	75.0	—	—	—	—	79.1	81.0	149.9	—	—	—
tc100	1,979	80.6	70.8	—	—	—	—	72.6	75.9	163.8	—	—	—

Note: Learned subdomains for CRIS are marked by * and generalized subdomains by +. Performance of HITEC is from the literature [13, 17]. Costs of our experiments are running times in seconds on a Sun SparcStation 10/512; costs of HITEC are running times in seconds on a Sun SparcStation SLC [16] (around four to six times slower than a Sun SparcStation 10/512).

Table 5.5 were based on ten runs of the underlying problem solver, implying that the computational cost is ten times of the average cost.

Table 5.6 summarizes the improvements of our learned and generalized HMs as compared to the published results of CRIS, GATEST, and HITEC. Each entry of the table shows the number of times our HM wins, ties, and loses in terms of fault coverages with respect to the method(s) in the first column. Our results show that our generalized HM based on CRIS as the problem solver is better than the original CRIS in 16 out of 21 circuits in terms of the maximum fault coverage, and 11 out of 21 circuits in terms of the average fault coverage. Further, our generalized HM based on GATEST as the problem solver is better than the original GATEST in 7 out of 19 circuits in terms of both the average and maximum fault coverages. Our results show that our generalization procedure can discover good HMs that work better than the original HMs.

Finally, we plot the distributions of symmetric fault coverages of our generalized HMs normalized with respect to average fault coverages of the original CRIS (Figure 5.5) and GATEST (Figure 5.6). These plots clearly demonstrate improvements over the original systems.

5.4.2 Heuristics for VLSI Placement and Routing

In our second application, we take TimberWolf [20] as our problem solver. This is a software package based on simulated annealing [8] to place and route various components (transistors, resistors, capacitors, wires, etc.) on a piece of silicon. Its goal is to minimize the chip area needed while satisfying constraints such as the number of layers of poly-silicon for routing and the maximum signal delay through the paths. Its operations can be divided into three steps: placement, global routing, and detailed routing.

The placement and routing problem is NP-hard; hence, heuristics are generally used. Simulated annealing (SA) used in TimberWolf is an efficient method to randomly search the space of possible placements. Although in theory SA converges asymptotically to the global optimum with probability one, the results generated in finite time are usually suboptimal. As a result, there is a trade-off between quality of a result and cost (or computational time) of obtaining it. In TimberWolf version 6.0, the version we have studied in this subsection, there are two parameters to control the running time (which indirectly control the quality of the result): *fast-n* and *slow-n*. The larger the *fast-n* is, the shorter time SA will run. In contrast, the larger the *slow-n* is, the longer time SA will run. Of course, only one of these parameters can be used in a single experiment to control the running time.

TimberWolf has six major components: *cost function, generate function, initial temperature, temperature decrement, equilibrium condition,* and *stopping criterion.* Many parameters in these components have been tuned manually. However, their settings are generally heuristic because we lack domain knowledge for setting them optimally. In Table 5.7 we list the parameters we have focused in this subsection. Our goal is to illustrate the power of our learning and generaliza-

Table 5.6 Summary of Results Comparing the Performance of our Generalized HMs with Respect to That of HITEC, CRIS, and GATEST

Our HM wins/ties/loses with respect to the following systems	CRIS generalized HM			GATEST generalized HM		
	Total	Max FC	Avg FC	Total	Max FC	Avg FC
HITEC	22	6, 2, 14	4, 0, 18	19	5+2, 2, 10	4+2, 1, 12
Original CRIS	21	16, 1, 4	11, 0, 10	18	18, 0, 0	17, 0, 1
Original GATEST	19	4, 3, 12	3, 0, 16	19	7+2, 7, 3	7+2, 8, 2
Both HITEC and CRIS	21	5, 2, 14	3, 0, 18	18	5+2, 1, 10	3+2, 1, 13
Both HITEC and GATEST	19	3, 3, 13	1, 0, 18	19	3+2, 4, 10	2+2, 2, 13
All HITEC, CRIS, and GATEST	18	2, 3, 13	1, 0, 17	18	3+2, 3, 10	1+2, 1, 14

Note: The first number in each entry shows the number of wins out of all applicable circuits, the second, the number of ties, and the third, the number of losses. A second number in the entry on wins indicates the number of circuits in which the test efficiency is already 100%. For these circuits, no further improvement is possible.

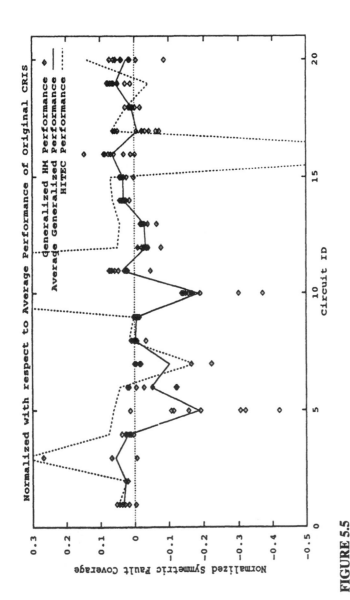

FIGURE 5.5

Distribution of normalized symmetric fault coverages of our generalized HM with respect to average fault coverages of the original CRIS on 20 benchmark circuits (s298, s344, s382, s386, s400, s444, s526, s641, s713, s820, s832, s1196, s1238, s1488, s1494, s1423, s5378, am2910, div16, and tc100 in that order).

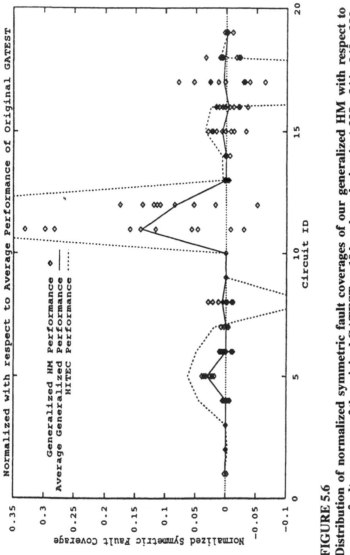

FIGURE 5.6
Distribution of normalized symmetric fault coverages of our generalized HM with respect to average fault coverages of the original GATEST on 19 benchmark circuits (s298, s344, s349, s382, s386, s400, s444, s526, s641, s713, s820, s832, s1196, s1238, s1423, s1488, s1494, s5378, and s35932 in that order).

Table 5.7 Parameters in TimberWolf (Version 6) Used in Our HM for Learning and Generalization

Parameter	Range	Step	Meaning	Default	Learned
P_1	0.1–2.5	0.1	Vertical path weight for estimating the cost function	1.0	0.1954624
P_2	0.33–2.0	0.1	Range limiter window change ratio	1.0	1.004637
P_3	18.0–28.0	1.0	High temperature finishing point	23.0	24.88345
P_4	71.0–91.0	1.0	Intermediate temperature finishing point	81.0	71.0
P_5	0.29–0.59	0.01	Critical ratio that determines acceptance probability	0.44	0.29
P_6	0.01–0.12	0.01	Temperature for controller turnoff	0.06	0.01

Table 5.8 Benchmark Circuits Used in Our Experiments

Cell name	Cells	Nets	Pins	Implicit feedthru
fract	124	163	454	0
s298	133	138	741	98
s420	211	233	1,488	112
primary1	766	1,172	5,534	0
struct	1,888	1,920	5,407	0
primary2	3,014	3,817	12,014	0
industrial1	2,271	2,597	—	—

From LayoutSynth92, *Int. Workshop on Layout Synthesis*, ftp site: mcnc.mcnc.org under /pub/benchmark, 1992.

tion procedures and to show improved quality and reduced cost for the placement and routing of large circuits, despite the fact that only small circuits were used in learning and in generalization.

In our experiments, we used seven benchmark circuits [10] whose specifications are shown in Table 5.8. Here, we have only studied the application of TimberWolf to standard-cell placement, though other kinds of placement (such as gate-array placement and macro/custom-cell placement) can be studied in a similar fashion. In our experiments, we used *fast-n* values of 1, 5, and 10, respectively. We first applied TEACHER to learn good HMs for circuits *s298* with *fast-n* of 1, *s420* with *fast-n* of 5, and *primary1* with *fast-n* of 10, each of which was taken as a learning subdomain. Each learning experiment involved 1000 applications of TimberWolf. Based on the best 30 HMs (10 from each subdomain), we applied our generalization procedure to obtain one generalized HM.

The default and generalized HMs are shown in Table 5.7. Table 5.9 compares

Table 5.9 Comparison of Cost-Quality between Our Generalized HM and the Default HM on Seven Benchmark Circuits and Three $fast\text{-}n$ Values over Ten Runs

Circuit	Performance measure	$fast\text{-}n = 10$		$fast\text{-}n = 5$		$fast\text{-}n = 1$	
		Default	Generalized	Default	Generalized	Default	Generalized
s298	Avg quality	655,706	0.019	0.008	0.040	0.020	0.045
	Max quality	640,668	0.044	0.038	0.052	0.032	0.058
	Avg cost	21.84	−0.062	0.454	0.412	3.525	3.282
s420	Avg quality	858,110	0.012	0.012	0.032	0.036	0.062
	Max quality	850,662	0.033	0.030	0.056	0.046	0.075
	Avg cost	30.69	−0.033	0.414	0.430	3.267	3.067
fract	Avg quality	87,524	0.061	0.054	0.049	0.099	0.168
	Max quality	77,248	0.166	0.162	0.152	0.183	0.229
	Avg cost	22.48	0.031	0.314	0.349	2.481	2.086
primary1	Avg quality	3,542,722	0.095	0.072	0.141	0.160	0.200
	Max quality	3,413,835	0.127	0.106	0.190	0.212	0.230
	Avg cost	227.10	−0.035	0.305	0.235	2.365	2.076
struct	Avg quality	2,029,258	0.405	0.225	0.622	0.658	1.000
	Max quality	1,894,847	0.528	0.293	0.755	0.781	1.119
	Avg cost	638.31	−0.414	0.538	0.280	2.497	2.179
primary2	Avg quality	17,944,576	0.100	0.150	0.225	0.273	0.315
	Max quality	17,111,600	0.173	0.205	0.282	0.315	0.337
	Avg cost	1,528.79	0.093	0.462	0.409	2.025	1.937
industry1	Avg quality	16,329,438	0.019	0.031	0.049	0.064	0.077
	Max quality	15,855,015	0.034	0.052	0.067	0.088	0.092
	Avg cost	1,321.97	0.109	0.200	0.165	1.987	1.702

Note: All performance values except the defaults with $fast\text{-}n=10$ are in the form of symmetric improvements with respect to the default performance values of $fast\text{-}n=10$. Percentage improvement in quality means percentage decrease in chip area.

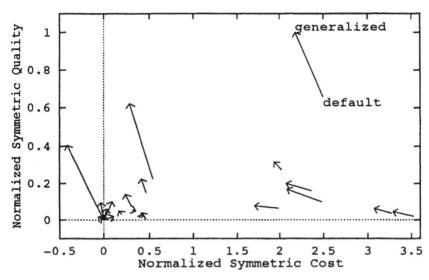

FIGURE 5.7
Comparison of average performance between the default and the generalized HMs.

the quality (average/maximum area of chip) and cost (average execution time) between the generalized HM and the default HM on all seven circuits with *fast-n* of 1, 5, and 10, respectively. (The cost for finding the minimum area is ten times the average cost.) These results are also plotted in Figure 5.7, where each arrow points from the average performance of the default HM to the average performance of the generalized HM.

Among the 21 test cases, the generalized HM has worse quality than that of the default in only one instance, and has worse cost in 5 out of 21 cases. Similarly, we see in Figure 5.7 that most of the arrows point in a left-upward direction, implying improved quality and reduced cost. Since we have only addressed a few parameters in TimberWolf, we expect to see more improvement as we learn other functions and parameters in TimberWolf.

5.4.3 Branch-and-Bound Search

A branch-and-bound search algorithm is a systematic method for traversing a search tree or search graph in order to find a solution that optimizes a given objective while satisfying the given constraints. It decomposes a problem into smaller subproblems and repeatedly decomposes them until a solution is found or infeasibility is proved. Each subproblem is represented by a node in the search tree/graph. The algorithm has four sets of HMs: (1) *selection HM* for selecting a search node for expansion based on a sequence of selection keys for ordering search nodes; (2) *decomposition HM* (or branching mechanism) for expanding a

Table 5.10   Generation of Test Cases for Learning and Generalization of Decomposition HMs in a Branch-and-Bound Search (Each Has 12 Subdomains)

Application	Subdomain attributes
VC	Connectivity of vertices is (0.05, 0.1, 0.15, 0.2, 0.25, 0.3, 0.35, 0.4, 0.45, 0.5, 0.55, or 0.6) Number of vertices is between 16 and 45
TSP	Distributions of 8–18 cities (uniformly distributed between 0–100 on both X and Y axes, uniformly distributed on one axis and normally distributed on another, or normally distributed on both axes) Graph connectivity of cities is (0.1, 0.2, 0.3, or 1.0)
KS	Range of both profits and weights is {(100–1000), (100–200), (100–105)} Variance of profit/weight ratio is (1.05, 1.5, 10, 100) 13–60 objects in the knapsack

search node into descendants using operators to expand (or transform) a search node into child nodes; (3) *pruning HM* for pruning inferior nodes in order to trim potentially poor subtrees; and (4) *termination HM* for determining when to stop.

In this subsection, we apply learning to find only new *decomposition HMs*; preliminary results on learning of selection and pruning HMs can be found in Reference [2]. We consider optimization search, which involves finding the optimal solution and proving its optimality.

We illustrate our method on three applications: traveling salesman problems on incompletely connected graphs mapped on a two-dimensional plane (TSP), vertex-cover problems (VC), and knapsack problems (KS). The second problem can be solved by a polynomial-time approximation algorithm with guaranteed performance deviations from optimal solutions, and the last can be solved by a pseudo-polynomial-time approximation algorithm. Hence, we expect that improvements due to learning are likely for the first two problems and not likely for the last. Table 5.10 shows the parameters used in generating a *test case* in each application. We further assume that all subdomains belong to one problem subspace.

The problem solver here is a branch-and-bound algorithm, and a test case is considered solved when its optimal solution is found. Note that the decomposition HM studied is only a component of the branch-and-bound algorithm. We use well-known decomposition HMs developed for these applications as our baseline HMs (see Table 5.12). The normalized cost of a candidate decomposition HM is defined in terms of its *average symmetric speedup* (see Equation 5.2), which is related to the number of nodes expanded by a branch-and-bound search using the baseline HM and that using the new HM. Note that we do not need to measure quality as both the new and existing HMs when applied in a branch-and-bound search look for the optimal solution.

In our experiments, we selected six subdomains in each application for learning. We performed learning in each of these subdomains using 1600 tests, selected the top five HMs in each subdomain, fully verified them on all the learned subdomains, and selected one final HM to be used across all the subdomains. Table 5.11 summarizes the learning, generalization, and validation results. In the learning results, we show the average symmetric speedup of the top HM learned in each subdomain and the normalized cost of learning, where the latter was computed as the ratio of the total CPU time for learning and the harmonic mean of the CPU times required by the baseline HM on test cases used in learning. The results show that a new HM learned specifically for a subdomain has around 1 to 35% improvement in its average symmetric speedups and 3000 to 16,000 times in learning costs.

Table 5.11 also shows the average symmetric speedups of the generalized HMs. We picked six subdomains randomly for learning. After learning and full verification of the five top HMs in each subdomain, we applied Equation 5.8 to identify one top HM to be used across all the 12 subdomains. Our results show that we have between 0 and 8% improvement in average symmetric speedups using the generalized HMs. Note that these results are worse than those obtained by learning, and that the baseline HM is the best HM for solving the knapsack problem.

The third part of Table 5.11 shows the average symmetric speedups when we validate the generalized HMs on larger test cases. These test cases generally require 10 to 50 times more nodes expanded than those used earlier. Surprisingly, our results show better improvement (9 to 23%). It is interesting to point out that 6 of the 12 subdomains with high degree of connectivity in the vertex-cover problem have slowdowns. This is a clear indication that these subdomains should be grouped in a different subspace and learned separately.

Table 5.12 shows the new decomposition HMs learned for the three applications. We list the variables that we fed to the learning system. In addition to these variables, we have also included constants that can be used by the heuristics generator. An example of such a constant is shown in the HM learned for the vertex-cover problem. This formula can be interpreted as using l as the primary key for deciding which node to include in the covered set. If the ls of two alternatives are different, then the remaining terms in the formula $(n - \Delta l)$ are insignificant. In contrast, when the ls are the same, then we use $n - \Delta l$ as a tie breaker.

Finally, Figure 5.8 plots the distribution of symmetric speedups of the generalized HM for VC with respect to the original HM using test cases in our generalization database. It shows the performance improvement of each individual test case with respect to that of the original HM. It further shows that performance is fairly evenly distributed above and below the average value without unnatural compression of ranges. This observation confirms that symmetric speedup is a proper normalization measure in this case. This plot also shows the difference in performance distribution across different subdomains, indicating the need to use a range-independent measure such as the probability of win.

Table 5.11 Results of Learning and Generalization for VC, TSP, and KS

Type	Applic-ation	Perfor-mance measure	Subdomain												Average
			1	2	3	4	5	6	7	8	9	10	11	12	
Learn-ing	VC	Sym-SU	0.000	0.011	0.041	0.000	0.044	0.022	0.008	0.013	0.000	0.000	0.000	0.000	0.012
		Cost	26,343.5	23,570.9	21,934.1	12,951.6	11,034.3	12,414.4	5,871.0	8,093.3	6,878.0	5,051.2	3,826.2	3,107.3	11,756.3
	TSP	Sym-SU	0.194	0.073	0.288	0.378	0.106	0.068	0.267	0.382	0.048	0.165	0.208	0.083	0.188
		Cost	2,846.6	1,543.9	2,077.7	2,207.7	2,314.9	1,865.6	1,889.9	1,847.5	2,509.7	1,947.0	1,445.4	1,958.8	2,037.9
	KS	Sym-SU	0.000	0.000	0.000	0.000	0.000	0.000	0.893	0.000	0.263	0.107	2.840	0.089	0.349
		Cost	25,707.7	32,587.9	9,671.6	26,408.1	24,903.6	22,309.1	3,648.1	7,943.1	8,114.7	6,476.2	772.9	10,684.4	14,935.6
Genera-lization	VC	Sym-SU	0.218	0.283*	0.031	0.068*	0.054	0.060*	0.017	0.049*	0.016	−0.000*	−0.011	0.028*	0.068
	TSP	Sym-SU	0.072*	0.004	0.082*	0.225	0.005*	0.061*	0.139	0.155	−0.010	0.054	0.090*	0.083*	0.080
	KS	Sym-SU	0.000*	0.000*	0.000	0.000	0.000	0.000*	0.000*	0.000	0.000*	0.000	0.000	0.000*	0.000
Valid-ation	VC	Sym-SU	0.070	0.638	0.241	0.078	0.073	0.020	−0.013	−0.004	−0.018	−0.000	−0.019	−0.010	0.088
	TSP	Sym-SU	0.417	0.036	0.144	0.155	0.131	0.364	1.161	0.101	0.108	0.008	0.022	0.131	0.231
	KS	Sym-SU	0.000	0.000	0.000	0.000	0.000	0.000	0.000	0.000	0.000	0.000	0.000	0.000	0.000

Note: In the results on generalization, numbers with * are the ones learned; only one common HM is generalized to all 12 subdomains.

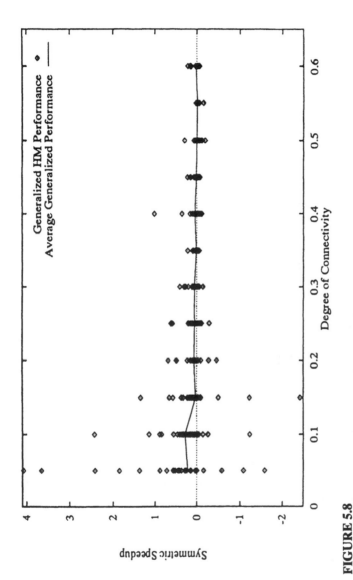

FIGURE 5.8
Distribution of performance values of the generalized decomposition HM for VC normalized on a case-by-case basis with respect to those of the original HM (using generalization test database).

Table 5.12 Original and Generalized Decomposition HMs Used in a Branch-and-Bound Search

Application	Variables used in constructing HMs	Original HM	Generalized HM
VC	$l =$ live degree of vertex (uncovered edges) $d =$ dead degree of vertex (covered edges) $n =$ average live degree of all neighbors of vertex $\Delta l =$ difference between l from parent node to current node	l	$1000\, l + n - \Delta l$ (l is the primary key, $n - \Delta l$ is the secondary key)
TSP	$c =$ length of current partial tour $m =$ min length to complete current partial tour $a =$ avg length to complete current partial tour $l =$ number of neighbor cities not yet visited $d =$ number of neighbor already visited	c	$m * c$
KS	$p, w =$ profit/weight of object $s =$ weight slack $=$ weight limit $-$ current weight $p_{max}, p_{min} =$ max/min profit of unselected objects $w_{max}, w_{min} =$ max/min weight of unselected objects	p/w	p/w

In short, our results show that reasonable improvements can be obtained by generalization of learned HMs. We anticipate further improvements by (1) learning and generalizing new pruning HMs in a depth-first search, (2) partitioning the problem space into a number of subspaces and learning a new HM for each, and (3) identifying attributes that help explain why one HM performs well in one subdomain but not in others.

5.5   Conclusions

In this chapter, we have studied automated generalization of performance-related heuristics for knowledge-lean applications. To summarize, we have derived the following results:

1. We have found inconsistencies in performance evaluation of heuristics due to multiple tests, multiple learning objectives, normalization, and changing behavior of heuristics across problem subdomains. We have proposed methods to cope with some of these anomalies.

2. We have studied methods to generalize learned heuristics to unlearned domains. To this end, we have proposed and evaluated a range-independent measure called probability of win for ranking heuristics in a problem subdomain. This allows heuristics across problem subdomains to be compared in a uniform manner. In case that there are trade-offs between cost and quality, our learning system will propose alternative heuristics showing such trade-offs.

3. We have found better heuristics for generating patterns in circuit testing, placement and routing of VLSI components, and branch-and-bound search. Due to space limitation, new heuristics for process mapping [7] and blind equalization in communication are not shown.

There are still several areas that we plan to study in the future.

1. One of the open problems that has not been studied is the identification of problem subdomains for learning and subspaces for generalization. Since such demarcation is generally vague and imprecise, we plan to apply fuzzy sets to help define subdomains and subspaces. Fuzzy logic can also help identify heuristics that can be generalized, especially when there are multiple objectives in the application.

2. We plan to study metrics for performance evaluation besides the average metric studied in this chapter. One such metric is the maximum metric that is useful when a heuristic method can be applied multiple times in order to generate better results at higher costs. This is also related to better generalization procedures that trade between improved quality and higher cost.

3. Finally, we plan to carry out learning and generalization on more applications. The merits of our system, of course, lie in finding better heuristics for real-world applications, which may involve many contradicting objectives. Our generalization procedure needs to be extended in order to cope with applications with multiple objectives.

Acknowledgments

The authors are grateful to the following people for providing us with application programs used in testing our generalization procedure: Daniel Saab for his CRIS system [17], Elizabeth Rudnick and Janak Patel for their GATEST system [16], and Lon-Chan Chu for his WISE system for evaluating branch-and-bound search [21].

This research was supported in part by National Science Foundation Grants MIP 92-10584 and MIP 88-10584 and National Aeronautics and Space Administration Grants NCC 2-481, NAG 1-613, and NGT 50743 (NASA Graduate Fellowship Program).

References

[1] Brglez, F., Bryan, D., and Kozminski, K., Combinatorial profiles of sequential benchmark circuits, *Int. Symp. Circuits Syst.,* p. 1929, May 1989.

[2] Chu, L.-C., Algorithms for Combinatorial Optimization in Real Time and Their Automated Refinement by Genetic Programming, Ph.D. thesis, Department of Electrical and Computer Engineering, University of Illinois, Urbana, May 1994.

[3] Edmonds, J. and Karp, R.M., Theoretical improvements in algorithmic efficiency for network flow problems, *J. ACM,* 19(2), 248, 1972.

[4] Gembicki, F.W., Vector Optimization for Control with Performance and Parameter Sensitivity Indices, Ph.D. thesis, Case Western Reserve University, Cleveland, OH, 1974.

[5] Goldberg, D.E. and Holland, J.H., Genetic algorithms and machine learning, in *Machine Learning,* Vol. 3, No. 2/3, Kluwer Academic Publishing, Boston, MA, October 1988, 95.

[6] Ieumwananonthachai, A., Aizawa, A. N., Schwartz, S. R., Wah, B. W., and Yan, J.C., Intelligent mapping of communicating processes in distributed computing systems, in *Proc. Supercomputing 91,* ACM/IEEE, Albuquerque, NM, November 1991, 512.

[7] Ieumwananonthachai, A., Aizawa, A., Schwartz, S. R., Wah, B. W., and Yan, J.C., Intelligent process mapping through systematic improvement of heuristics, *J. Parallel Distributed Comput.*, 15, 118, 1992.

[8] Kirkpatrick, S., Gelatt, C. D., Jr., and Vecchi, M. P., Optimization by simulated annealing, *Science*, 220(4598), 671, 1983.

[9] Koza, J. R., *Genetic Programming*, MIT Press, Cambridge, MA, 1992.

[10] LayoutSynth92, *Workshop on Layout Synthesis*, ftp site: mcnc.mcnc.org under/pub/benchmark, 1992.

[11] Mehra, P. and Wah, B. W., *Load Balancing: An Automated Learning Approach*, World Scientific Publishing, Singapore, 1995.

[12] Newell, A., Shaw, J. C., and Simon, H. A., Programming the logic theory machine, *Proc. 1957 West. Joint Comput. Conf.*, IRE, 230, 1957.

[13] Niermann, T.M. and Patel, J. H., HITEC: a test generation package for sequential circuits, *Eur. Design Automation Conf.*, p.214, 1991.

[14] Pearl, J., *Heuristics–Intelligent Search Strategies for Computer Problem Solving*, Addison-Wesley, Reading, MA, 1984.

[15] Ramsey, C. L. and Grefenstette, J.J., Case-based initialization of genetic algorithms, in *Proc. 5th Int. Conf. on Genetic Algorithms*, pp. 84–91, International Society for Genetic Algorithms, Champaign, IL, June 1993.

[16] Rudnick, E.M., Patel, J.H., Greenstein, G.S., and Niermann, T.M., Sequential circuit test generation in a genetic algorithm framework, in *Proc. Design Automation Conf.*, ACM/IEEE, Los Alamitos, CA, June 1994.

[17] Saab, D.G., Saab, Y.G., and Abraham, J.A., CRIS: a test cultivation program for sequential VLSI circuits, in *Proc. Int. Conf. on Computer Aided Design*, IEEE, Santa Clara, CA, November 8–12, 1992, 216.

[18] Schwartz, S.R. and Wah, B.W., Automated parameter tuning in stereo vision under time constraints, in *Proc. Int. Conf. on Tools for Artificial Intelligence*, IEEE, Los Alamitos, CA, November 1992, 162.

[19] Sechen, C. and Sangiovanni-Vicentelli, A., The TimberWolf placement and routing package, *J. Solid State Circuits*, 20(2), 510, 1985.

[20] Sechen, C., *VLSI Placement and Global Routing Using Simulated Annealing*, Kluwer Academic Publishers, Boston, MA, 1988.

[21] Wah, B.W. and Chu, L.-C., Combinatorial search algorithms with meta-control: modeling and implementations, *Int. J. Artif. Intelligence Tools,* 1(3), 369, 1992.

[22] Wah, B.W., Population-based learning: a new method for learning from examples under resource constraints, *Trans. Knowledge Data Eng.,* 4(5), 454, 1992.

[23] Wah, B.W., Ieumwananonthachai, A., Chu, L.-C., and Aizawa, A., Genetics-based learning of new heuristics: rational scheduling of experiments and generalization, *IEEE Trans. Knowledge Data Eng.,* 7(5), 763–785, October 1995.

6

Genetic Algorithm-Based Pattern Classification: Relationship with Bayes Classifier

Chivukula A. Murthy, Sanghamitra Bandyopadhyay, and Sankar K. Pal

Abstract   In this chapter, an investigation is carried out to formulate theoretical results regarding the behavior of a genetic algorithm-based pattern classification methodology, for an infinitely large number of sample points n, in an N dimensional space $\mathcal{R}^N$. It is shown that for $n \to \infty$, and for a sufficiently large number of iterations, the performance of this classifier approaches that of the Bayes classifier. Experimental results, for a triangular distribution of points, are also included that conform to this claim.

6.1 Introduction

Genetic algorithm (GA) [1, 2] is a stochastic process performing search over a complex and multimodal space. It is a randomized method in that it utilizes domain specific knowledge, in the form of objective function, to perform a directed random search. GAs are gradually finding applications in a wide variety of fields like VLSI design, machine learning, job shop scheduling, etc. [3].

Recently, an application of GA has been reported in the area of pattern classification in N dimensional space, where it has been used for the placement of H hyperplanes (constituting the decision boundary) in the feature space such that they provide the minimum misclassification of the sample points [4]. The methodology described in Reference [4] requires an *a priori* estimation of H. Due to the difficulty in acquiring such a knowledge, H is frequently overestimated, which may lead to the presence of redundant hyperplanes in the decision boundary. Thus a scheme for the deletion of these redundant hyperplanes has also been formulated which is described in Reference [5].

Bayes maximum likelihood classifier [6] is one of the most widely used statistical pattern classifiers which computes the total expected loss of assigning a sample point to the various classes. It then performs a classification such that this expected loss is minimized. It is known in the literature [6] that Bayes classifier provides the optimal performance from the standpoint of error probabilities in a statistical framework. This classifier is generally accepted as a standard against which the performance of other classifiers may be compared. A comparison of the recognition scores obtained by both the GA-based classifier [4] and Bayes classifier [6] has been provided in References [4] and [5] for both nonoverlapping, concave, and overlapping data sets.

In this chapter we attempt to show theoretically that the decision surfaces generated by the aforesaid GA-based classifier will approach the Bayes decision surfaces for a large number of training sample points (n) and will consequently provide the optimal decision boundary in terms of the number of misclassified samples. As already stated, in view of the difficulty in assessing H *a priori*, it is frequently overestimated. We also provide a proof that such an overestimation, coupled with the scheme for deletion of redundant hyperplanes, will provide the optimal value of H for a large number of sample points when there is no confusion in the Bayes classification rule itself. In situations where the Bayes decision regions are not unique, estimation of the optimal value of H, in spite of the scheme for deleting the redundant hyperplanes, may not be possible, although the classifier will still be optimal in terms of the number of misclassified points.

The chapter is organized as follows: Section 6.2 gives a brief outline of the GA-based classifier introduced in References [4] and [5]. Section 6.3 provides the theoretical treatment to find a relationship between this classifier and the Bayes classifier in terms of classification accuracy. A critical discussion of the proof is also given in the same section. This is followed by a description of the process of redundancy elimination along with the associated proof regarding the optimal performance, and the estimation of the optimal value of H in Section 6.4. Section 6.5 contains the experimental results and their analysis. Finally, the conclusions are presented in Section 6.6.

6.2   The GA-Based Classifier: A Brief Outline

In genetic algorithms, an individual solution of the problem is encoded in the form of strings or chromosomes. An objective function is associated with each string which provides a mapping from the chromosomal space to the problem space.

An initial population is created by randomly selecting a fixed number of strings. Biologically inspired operators like *crossover* and *mutation* are then applied over them to yield a new pool of strings. Based on the principle of *survival of the fittest*, a few among them are selected (some more than once) to form a new population. This cycle of crossover, mutation, and selection is repeated for a number of times until a termination condition is achieved. The best string seen up to the last generation generally provides the solution to the problem under consideration.

In the realm of pattern classification using a fixed number (H) of hyperplanes, the chromosomes are representatives of a set of such hyperplanes. The associated objective function is the number of points misclassified (*miss*) by the set of hyperplanes encoded in the string. The fitness function that the GA attempts to maximize is the number of points correctly classified by a string, i.e., $fit = n - miss$, where n is the total number of training sample points.

6.2.1 String Representation

From elementary geometry, the equation of a hyperplane in N dimensional space $(X_1 - X_2 - \cdots - X_N)$ is given by

$$x_N \cos \alpha_{N-1} + \beta_{N-1} \sin \alpha_{N-1} = d$$

where

$$\beta_{N-1} = x_{N-1} \cos \alpha_{N-2} + \beta_{N-2} \sin \alpha_{N-2}$$

$$\beta_{N-2} = x_{N-2} \cos \alpha_{N-3} + \beta_{N-3} \sin \alpha_{N-3}$$

$$\vdots$$

$$\beta_1 = x_1 \cos \alpha_0 + \beta_0 \sin \alpha_0$$

The various parameters are as follows :

$(x_1, x_2, \ldots, x_N)$:	a point on the hyperplane
α_{N-1}:	the angle that the unit normal to the hyperplane makes with the X_N axis
α_{N-2}:	the angle that the projection of the normal in the $X_1 - X_2 - \cdots - X_{N-1}$ space makes with the X_{N-1} axis
$\vdots$	
α_1:	the angle that the projection of the normal in the $(X_1 - X_2)$ plane makes with the X_2 axis
α_0:	the angle that the projection of the normal in the (X_1) plane makes with the X_1 axis $= 0$
d:	the perpendicular distance of the hyperplane from the origin

Thus, the N tuple $< \alpha_1, \alpha_2, \ldots, \alpha_{N-1}, d >$ specifies a hyperplane in N dimensional space.

Each angle α_j, $j = 1, 2, \ldots, N - 1$ is allowed to vary in the range of 0 to 2π, with the precision defined by the number of bits used for representing an angle. In order to specify d, the hyper rectangle enclosing the sample points is considered. For $N - 1$ angles, $\alpha_1, \ldots, \alpha_{N-1}$ (i.e., for a given orientation), the hyperplane passing through one of the vertices of the hyper rectangle and having a minimum distance, d_{min}, from the origin is specified as the base hyperplane. Let $diag$ be the length of the diagonal of the hyper rectangle and let the bits used for specifying d be capable of generating values, say *offset*, in the range [0,$diag$]. (The number of bits again controls the precision of d.) Then $d = d_{min}+offset$.

6.2.2 Genetic Operators

A set of H hyperplanes, each comprising $N-1$ angle variables and d, is encoded in a single chromosome. A chromosome is thus represented by the H tuple $< \mathcal{H}_1, \mathcal{H}_2, \ldots, \mathcal{H}_H >$ where each $\mathcal{H}_i, i = 1, 2, \ldots, H$ is represented by an N tuple $< \alpha_1, \alpha_2, \ldots, \alpha_{N-1}, d >$.

In order to initialize the population, Pop binary strings are generated randomly. The fitness of a string is defined as the number of points properly classified by the H hyperplanes encoded in it. For computing the fitness of a string, the parameters $\alpha_1, \alpha_2, \ldots, \alpha_{N-1}$ and d, corresponding to each hyperplane encoded in it, are extracted. These are used to determine the region in which each training pattern point lies. A region is said to provide the demarcation for class i if the maximum number of points that lie in this region are from class i. Any tie is resolved arbitrarily. All other points in this region are considered to be misclassified. The misclassifications for all the regions are summed up to provide the number of misclassifications, *miss*, for the entire string. Its fitness is then defined to be $(n\text{-}miss)$.

Roulette wheel selection is adopted to implement the *proportional selection strategy*. Elitism is incorporated by replacing the worst string of the current generation with the best string seen up to the last generation. *Single-point* crossover is applied with a fixed crossover probability value of 0.8. The mutation operation is performed on a bit-by-bit basis for a mutation probability value which is initially high, then decreased gradually to a prespecified minimum value and then increased again in the later stages of the algorithm. This ensures that in the initial stage, when the algorithm has very little knowledge about the search domain, it performs a random search through the feature space. This randomness is gradually decreased with the passing of generations so that now the algorithm performs a detailed search in the vicinity of promising solutions obtained so far. In spite of this, the algorithm may still get stuck at a local optima. This problem is overcome by increasing the mutation probability to a high value, thereby making the search more random once again. The algorithm is terminated if the population contains at least one string with zero misclassification of points and there is no significant improvement in the average fitness of the population over subsequent generations. Otherwise, the algorithm is executed for a fixed number of generations. ♣

Experimental results regarding the satisfactory performance of the algorithm for a variety of data sets, with both overlapping and nonoverlapping, nonlinearly separable classes, are provided in References [4] and [5]. It is our aim in this chapter to provide a theoretical basis of the proposed algorithm, and to prove that for $n \to \infty$ and for a sufficiently large number of iterations, the proposed algorithm will try to approach the Bayes classifier.

6.3   Relationship with Bayes Classifier

Let there be k classes $C_1, C_2, \ldots, C_k$ with *a priori* probabilities $P_1, P_2, \ldots, P_k$ and class conditional densities $p_1(x), p_2(x), \ldots, p_k(x)$. Let the mixture density be

$$p(x) = \sum_{i=1}^{k} P_i \, p_i(x).$$

Let $X_1, X_2, \ldots, X_n, \ldots$ be independent and identically distributed (i.i.d) N dimensional random vectors with density $p(x)$. This indicates that there is a probability space $(\Omega, \mathcal{F}, Q)$, where $\mathcal{F}$ is a σ field of subsets of Ω, Q is a probability measure on $\mathcal{F}$, and

$$X_i : (\Omega, \mathcal{F}, Q) \longrightarrow \left[\mathcal{R}^N, B(\mathcal{R}^N), P \right], \ \forall i = 1, 2, \ldots$$

such that

$$P(A) = Q \left[X_i^{-1}(A) \right]$$

$$= \int_A p(x)dx$$

$\forall A \in B(\mathcal{R}^N)$ and $\forall i = 1, 2, \ldots.$
    Here $B(\mathcal{R}^N)$ is the Borel σ field of $\mathcal{R}^N$.
    Let us define $\mathcal{E}$ such that

$$\mathcal{E} = \{E : E = (S_1, S_2, \ldots, S_k), S_i \subseteq \mathcal{R}^N,$$

$$\forall i = 1, \ldots, k, \bigcup_{i=1}^{k} S_i = \mathcal{R}^N, S_i \bigcap S_j = \emptyset, \forall i \neq j\}.$$

$\mathcal{E}$ provides the set of all partitions of $\mathcal{R}^N$ into k sets as well as their permutations, i.e.,

$$E_1 = (S_1, S_2, S_3 \ldots, S_k) \in \mathcal{E}$$

$$E_2 = (S_2, S_1, S_3, \ldots, S_k) \in \mathcal{E}$$

then $E_1 \neq E_2$. Note that for $E = (S_{i_1}, S_{i_2}, \ldots, S_{i_k})$ implies that each S_{i_j}, $1 \leq j \leq k$, is the region corresponding to class j.

    Let $E_0 = (S_{01}, S_{02}, \ldots, S_{0k}) \in \mathcal{E}$ be such that each S_{0i} is the region corresponding to the class C_i in $\mathcal{R}^N$ and these are obtained by using Bayes decision rule. Then

$$a = \sum_{i=1}^{k} P_i \int_{S_{0i}^c} p_i(x) \leq \sum_{i=1}^{k} P_i \int_{S_{1i}^c} p_i(x)$$

$\forall E_1 = (S_{11}, S_{12}, \ldots, S_{1k}) \in \mathcal{E}$. Here a is the error probability obtained using the Bayes decision rule.

    It is known from the literature that such an E_0 exists and it belongs to $\mathcal{E}$, because Bayes decision rule provides an optimal partition of $\mathcal{R}^N$ and for every such $E_1 = (S_{11}, S_{12}, \ldots, S_{1k}) \in \mathcal{E}, \sum_{i=1}^{k} P_i \int_{S_{1i}^c} p_i(x)$ provides the error probability for $E_1 \in \mathcal{E}$.

Note that E_0 need not be unique.

    Assumptions: Let H be a positive integer and let there exist H hyperplanes in $\mathcal{R}^N$ which can provide the regions $S_{01}, S_{02}, \ldots, S_{0k}$. Let H be known *a priori*. Let the algorithm for generation of class boundaries using H hyperplanes be allowed to execute for a sufficiently large number of iterations. It is known in the literature [7] that as the number of iterations goes toward infinity, an elitist model of GA will certainly provide the optimal string.

    Let $\mathcal{A} = \{A : A$ is a set consisting of H hyperplanes in $\mathcal{R}^N\}$. Let $A_0 \in \mathcal{A}$ be such that it provides the regions $S_{01}, S_{02}, \ldots, S_{0k}$ in $\mathcal{R}^N$. Note that each

$A \in \mathcal{A}$ generates several elements of $\mathcal{E}$. Let $\mathcal{E}_A \subseteq \mathcal{E}$ denote all possible $E = (S_1, S_2, \ldots, S_k) \in \mathcal{E}$ that can be generated from A.

Let G	$= \bigcup_{A \in \mathcal{A}} \mathcal{E}_A$	
Let $Z_{iE}(\omega)$	$= 1$ if $X_i(\omega)$ is misclassified when E is used as a decision rule where $E \in G, \forall \omega \in \Omega.$	
	$= 0$ otherwise.	
Let $f_{nE}(\omega)$	$= \frac{1}{n} \sum_{i=1}^{n} Z_{iE}(\omega),$ when $E \in G$ is used as a decision rule.	
Let $f_n(\omega)$	$= \text{Inf} \{ f_{nE}(\omega) : E \in G \}.$	

It is to be noted that the pattern classification algorithm mentioned in Section 6.2 uses $n \times f_{nE}(\omega)$, the total number of misclassified samples, as the objective function which it attempts to minimize. This is equivalent to searching for a suitable E such that the term $f_{nE}(\omega)$ is minimized, i.e., for which $f_{nE}(\omega) = f_n(\omega)$. As already mentioned, it is known that for infinitely many iterations the elitist model of GAs will certainly be able to obtain such an E.

THEOREM 6.1:

For sufficiently large n, $f_n(\omega) \not> a$, i.e., $f_n(\omega)$ cannot be greater than a, almost everywhere, where $a = \sum_{i=1}^{k} P_i \int_{S_{0i}^c} p_i(x).$

PROOF   Let $Y_i(\omega) = 1$ if $X_l(\omega)$ is misclassified according to Bayes rule $\forall \omega \in \Omega$.
$\qquad\qquad\qquad = 0$ otherwise.

Note that $Y_1, Y_2, \ldots, Y_n, \ldots$ are i.i.d random variables. Now

$$Prob(Y_i = 1) = \sum_{j=1}^{k} Prob(Y_i = 1/X_i \text{ is in } C_j) P(X_i \text{ is in } C_j)$$

$$= \sum_{j=1}^{k} P_j Prob(\omega : X_i(\omega) \in S_{0j}^c \text{ given that } \omega \in C_j)$$

$$= \sum_{j=1}^{k} P_j \int_{S_{0j}^c} p_j(x)dx = a.$$

Hence the expectation of Y_i, $E(Y_i)$ is given by

$$E(Y_i) = a, \quad \forall i.$$

Then by using Strong Law of Large Numbers [8], $\frac{1}{n}\sum_{i=1}^{n}Y_i \longrightarrow a$ almost everywhere, i.e.,

$$P(\omega : \frac{1}{n}\sum_{i=1}^{n}Y_i(\omega) \nrightarrow a) = 0.$$

Let $B = \{\omega : \frac{1}{n}\sum_{i=1}^{n}Y_i(\omega) \longrightarrow a\} \subseteq \Omega$.

Obviously $f_n(\omega) \leq \frac{1}{n}\sum_{i=1}^{n}Y_i(\omega)$, $\forall n$ and $\forall \omega$, since the set of regions $(S_{01}, S_{02}, \ldots, S_{0k})$ obtained by the Bayes decision rule is also provided by some $A \in \mathcal{A}$ and consequently it will be included in G. Note that $0 \leq f_n(\omega) \leq 1$, $\forall n$ and $\forall \omega$. Let $\omega \in B$. For every $\omega \in B$, $U(\omega) = \{f_n(\omega); n = 1, 2, \ldots\}$ is a bounded, infinite set. Then by Bolzano-Weierstrass theorem [9], there exists an accumulation point of $U(\omega)$. Let $y = \text{Sup}\{y_0 : y_0 \text{ is an accumulation point}$ of $U(\omega)\}$. From elementary mathematical analysis we can conclude that $y \leq a$, since $\frac{1}{n}\sum_{i=1}^{n}Y_i(\omega) \longrightarrow a$ almost everywhere and $f_n(\omega) \leq \frac{1}{n}\sum_{i=1}^{n}Y_i(\omega)$. Thus it is proved that for sufficiently large n, $f_n(\omega)$ cannot be greater than a almost everywhere.

Note:

- The proof is not typical of the GA-based classifier. It will hold for any other classifier where the criterion is to reduce the number of misclassified sample points.

- Instead of hyperplanes, any other higher order surfaces could have been used for approximating the decision boundaries. It would lead to only minor modifications to the proof presented earlier, with the basic inference still holding good.

- The proof establishes that the number of points misclassified by the GA-based classifier will always be less than or equal to the number of points misclassified by Bayes decision rule, i.e., $f_n(\omega) \leq a$. However, the fact that $f_n(\omega) < a$ is true for only a finite number of sample points. This is due to the fact that a small number of identical sample points can be generated by different statistical distributions. Consequently, each distribution will result in different error probabilities of the Bayes classifier. The proposed GA classifier, on the other hand, will always find the decision surface yielding the smallest number of misclassified points, irrespective of their statistical properties.
  As the number of points increases, the number of possible distributions that can produce them decreases. In the limiting case when $n \to \infty$, only one distribution can possibly produce all the sample points, and Bayes

classifier designed over this distribution will provide the optimal decision boundary. The GA-based algorithm, in that case, will also yield a decision boundary which is same as the Bayes decision boundary. This fact has been borne out by the experimental results given in Section 6.5 that for an increased number of sample points, the decision surface provided by the GA classifier indeed approximates the Bayes decision boundary more closely. ♣

An important assumption to the proof of the theorem is that the value of H is known *a priori*. Since this knowledge is very difficult to acquire, H is frequently overestimated, leading to the presence of redundant hyperplanes in the resulting decision boundary. A scheme for the elimination of these redundant hyperplanes is described in Reference [5]. The scheme in Reference [5] may not eliminate all possible redundant hyperplanes since it is dependent on the order of testing the hyperplanes. Thus we describe a modified redundancy elimination procedure here, which guarantees optimality in terms of the number of nonredundant hyperplanes with respect to the given set of hyperplanes. In the following section we provide a critical analysis regarding its effectiveness in estimating the optimal H.

6.4   Optimum H and Bayes Decision Regions

It is to be noted that since H_1 hyperplanes provide a maximum of 2^{H_1} regions in the feature space, for a k class problem, k must be $\leq 2^{H_1}$; or $H_1 \geq \log_2 k$. A hyperplane is considered to be redundant if its removal has no effect on the recognition score of the classifier for the given training data set. In order to arrive at the optimal number of hyperplanes, all possible combinations of the H_1 hyperplanes must be considered while implementing redundancy removal. For each such combination, the first hyperplane is removed and it is tested whether the remaining hyperplanes can successfully classify all the patterns. If so, then this hyperplane is considered to be absent, and the test is repeated for the next hyperplane in the combination.

Obviously, testing all possible combinations results in an algorithm with exponential complexity in H_1. A branch-and-bound technique is adopted where search within a combination is discontinued (before considering all the H_1 hyperplanes) if the number of hyperplanes found to be nonredundant so far is greater than or equal to the number of hyperplanes declared to be nonredundant by some earlier combination. The complexity may be reduced further by terminating the algorithm if the combination being tested provides a set of $\lceil \log_2 k \rceil$ nonredundant hyperplanes, this being the minimum number of hyperplanes that can possibly be required for providing the k regions.♣

In the following discussion, we attempt to find a relationship between H and H_1.

6.4.1   Relationship between H and H_1

Let H be the optimal value of the number of hyperplanes required to model the Bayes decision boundary for a k class problem. Then any number $H_1 \geq H$ of hyperplanes will be sufficient to provide the required boundary, an essential assumption of our proof. Let us start with H_1 hyperplanes, an overestimation of H. As a consequence of the proof given in Section 6.3, the GA-based classifier will surely result in the decision regions with average misclassification $= f_n(\omega)$ which is less than or equal to a almost everywhere for $n \to \infty$, for the H_1 hyperplanes.

Let the method of removal of redundant hyperplanes find only H_2 of these H_1 hyperplanes to be essential to provide the decision surfaces for separating the k classes. Note that the process of elimination of redundant hyperplanes does not affect the recognition capability of the classifier. In other words, it does not alter the effective decision boundary obtained for the n points. Consequently, H_2 hyperplanes (i.e., those obtained after elimination of redundant hyperplanes) will also provide the same effective decision surface as that provided by the H_1 hyperplanes. Since H hyperplanes is the optimal value, H_2 cannot be less than H. It is now our task to ascertain whether $H_2 > H$ or $H_2 = H$.

In order to determine the relationship between H_2 and H, we must consider the following situations:

1. The number of partitions which provide the Bayes error probability is exactly one. Since this partition, formed from H hyperplanes, provides the Bayes error probability, which is known to be optimal, hence for $n \to \infty$, the regions provided by the H_2 hyperplanes must be exactly the same as the regions provided by the H hyperplanes. Thus H_2 must be the same as H for larger values of n.

2. On the contrary, the number of partitions that provide the Bayes error probability may be greater than one. To elaborately explain this point, let us consider a three-class problem.

Let the *a priori* probabilities for the three classes be P_1, P_2, and P_3 and the class conditional densities be $p_1(x)$, $p_2(x)$, and $p_3(x)$. Let the regions associated with the three classes be Ω_1, Ω_2, and Ω_3 such that $\Omega_i \bigcap \Omega_j = \emptyset$, $\forall i \neq j$, and $\Omega_1 \bigcup \Omega_2 \bigcup \Omega_3 = \Omega$. Then the error probability e is given by

$$e = \sum_{i=1}^{3} \int_{\Omega_i^c} P_i p_i(x)$$

$$= \int_{\Omega_2 \bigcup \Omega_3} P_1 p_1(x) + \int_{\Omega_1 \bigcup \Omega_3} P_2 p_2(x) + \int_{\Omega_1 \bigcup \Omega_2} P_3 p_3(x)$$

$$= \int_{\Omega_1} P_1 p_1(x) - \int_{\Omega_1} P_1 p_1(x) + \int_{\Omega_2 \bigcup \Omega_3} P_1 p_1(x) + \int_{\Omega_1 \bigcup \Omega_3} P_2 p_2(x) +$$

$$\int_{\Omega_1 \bigcup \Omega_2} P_3 p_3(x)$$

$$= P_1 - \int_{\Omega_1} P_1 p_1(x) + \int_{\Omega_1 \bigcup \Omega_3} P_2 p_2(x) + \int_{\Omega_1 \bigcup \Omega_2} P_3 p_3(x)$$

$$= P_1 + P_2 - \int_{\Omega_1} P_1 p_1(x) - \int_{\Omega_2} P_2 p_2(x) + \int_{\Omega_1 \bigcup \Omega_2} P_3 p_3(x)$$

$$= P_1 + P_2 + \int_{\Omega_1} \{P_3 p_3(x) - P_1 p_1(x)\} + \int_{\Omega_2} \{P_3 p_3(x) - P_2 p_2(x)\}$$

Similarly

$$e = P_2 + P_3 + \int_{\Omega_2} \{P_1 p_1(x) - P_2 p_2(x)\} + \int_{\Omega_3} \{P_1 p_1(x) - P_3 p_3(x)\}$$

and

$$e = P_3 + P_1 + \int_{\Omega_3} \{P_2 p_2(x) - P_3 p_3(x)\} + \int_{\Omega_1} \{P_2 p_2(x) - P_1 p_1(x)\}$$

Summing up, we get

$$3e = 2 + \int_{\Omega_1} \{[P_3 p_3(x) - P_1 p_1(x)] + [P_2 p_2(x) - P_1 p_1(x)]\} +$$

$$\int_{\Omega_2} \{[P_3 p_3(x) - P_2 p_2(x)] + [P_1 p_1(x) - P_2 p_2(x)]\} +$$

$$\int_{\Omega_3} \{[P_1 p_1(x) - P_3 p_3(x)] + [P_2 p_2(x) - P_3 p_3(x)]\}$$

Let γ be used to represent the term

$$\gamma = \int_{\Omega_1} \{[P_3 p_3(x) - P_1 p_1(x)] + [P_2 p_2(x) - P_1 p_1(x)]\} +$$

$$\int_{\Omega_2} \{[P_3 p_3(x) - P_2 p_2(x)] + [P_1 p_1(x) - P_2 p_2(x)]\} +$$

$$\int_{\Omega_3} \{[P_1 p_1(x) - P_3 p_3(x)] + [P_2 p_2(x) - P_3 p_3(x)]\}$$

Therefore $3e = 2 + \gamma$.

Bayes classifier classifies a point x to the class which minimizes e, i.e., which in effect minimizes γ. Accordingly, in order to classify a point x to one of the three classes the following cases may arise:

1. $\{x : P_1 p_1(x) > P_2 p_2(x) > P_3 p_3(x)\}$: classify to class 1.
2. $\{x : P_1 p_1(x) > P_2 p_2(x) = P_3 p_3(x)\}$: classify to class 1.
3. $\{x : P_1 p_1(x) > P_3 p_3(x) > P_2 p_2(x)\}$: classify to class 1.
4. $\{x : P_1 p_1(x) = P_3 p_3(x) > P_2 p_2(x)\}$: classify to class 1 or 3.
5. $\{x : P_3 p_3(x) > P_1 p_1(x) > P_2 p_2(x)\}$: classify to class 3.
6. $\{x : P_3 p_3(x) > P_1 p_1(x) = P_2 p_2(x)\}$: classify to class 3.
7. $\{x : P_3 p_3(x) > P_2 p_2(x) > P_1 p_1(x)\}$: classify to class 3.
8. $\{x : P_3 p_3(x) = P_2 p_2(x) > P_1 p_1(x)\}$: classify to class 3 or 2.
9. $\{x : P_2 p_2(x) > P_3 p_3(x) > P_1 p_1(x)\}$: classify to class 2.
10. $\{x : P_2 p_2(x) > P_3 p_3(x) = P_1 p_1(x)\}$: classify to class 2.
11. $\{x : P_2 p_2(x) > P_1 p_1(x) > P_3 p_3(x)\}$: classify to class 2.
12. $\{x : P_2 p_2(x) = P_1 p_1(x) > P_3 p_3(x)\}$: classify to class 2 or 1.
13. $\{x : P_1 p_1(x) = P_2 p_2(x) = P_3 p_3(x)\}$: classify to any of
    the three classes.

As is obvious from the previous discussion, regions represented by cases 4, 8, 12, and 13 do not have unique classification associated with them. These regions may be included in more than one class, while still providing the least error probability. Thus, there is a confusion inherent in the problem itself which gives rise to nonunique but optimal classifications. In these cases, the regions provided by the H_2 hyperplanes may not be identical to the ones provided by the H hyperplanes. Consequently, H_2 can be greater than H for such situations, although the classifier still provides an average error $= f_n(\omega)$.

6.5   Experimental Results

In Section 6.3, a theoretical study has been made regarding the claim that for a large value of n, the performance of the GA-based classifier will try to approach that of the Bayes classifier, which is known to be optimal in a statistical framework. Empirical results are provided in this section, which are seen to conform to the above-mentioned statement. The performance of the GA-based classifier is compared with that of the Bayes classifier as well as the k-NN classifier (for k = 1).

The two-dimensional $(X - Y)$ data set used for the experiments is generated artificially, using a triangular distribution of the form shown in Figure 6.1 for the two classes, 1 and 2. The range for class 1 is $[0,2] \times [0,2]$ and that for class 2 is $[1,3] \times [0,2]$ with the peaks at $(1,1)$ and $(2,1)$ for the two classes, respectively. Figure 6.1 shows the distribution along the X axis since only this axis has discriminatory capability. The distribution along the X axis may be formally quantified as

$$f_1(x) = \quad 0 \quad for \; x \leq 0$$

$$f_1(x) = \quad x \quad for \; 0 < x \leq 1$$

$$f_1(x) = 2 - x \; for \; 1 < x \leq 2$$

$$f_1(x) = \quad 0 \quad for \; x > 2$$

for class 1. Similarly for class 2

$$f_2(x) = \quad 0 \quad for \; x \leq 1$$

$$f_2(x) = x - 1 \; for \; 1 < x \leq 2$$

$$f_2(x) = 3 - x \; for \; 2 < x \leq 3$$

$$f_2(x) = \quad 0 \quad for \; x > 3.$$

The two classes are assigned *a priori* probabilities $P_1 = 0.4$ (for class 1) and $P_2 = 0.6$ (for class 2). Using elementary mathematics, we can show that Bayes classifier will classify a point to class 1 if its X coordinate is less than $1 + P_1$, which signifies that the Bayes decision boundary is given by $x = 1 + P_1$.

The experiments are performed for a varying number of data points n generated by the mentioned triangular distribution. The different values of n considered are 100, 200, 500, 1000, and 1500. The results are presented in Table 6.1 where it is seen that the GA-based classifier performs consistently better than Bayes classifier. Although the performance of the proposed classifier is better than that of the Bayes classifier by a significant margin for $n = 100$ and 200, the difference gradually decreases for a larger value of n. Indeed, for $n = 1500$, the results of the two are seen to be quite close. It is also observed that the recognition score for class 1 is consistently lower than that for class 2 which has a higher *a priori* probability

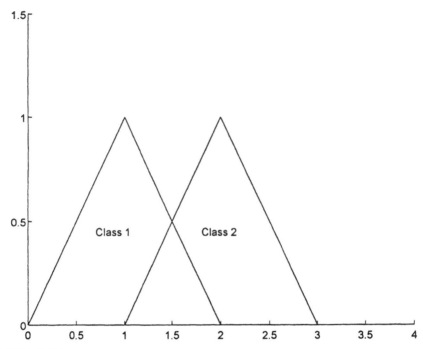

FIGURE 6.1
Triangular distribution for the two classes.

(= 0.6). The results presented in Table 6.1 are the average values computed over different data sets for the same *n* and over different initial population for the GA-based classifier.

As expected, the k-NN classifier performs poorly in each case. These results are included as a typical illustration of comparison.

Figures 6.2 to 6.4 show the decision boundary obtained by the proposed classifier and Bayes classifier for *n* = 100, 1000, and 1500, respectively. In Figure 6.2 the Bayes boundary is seen to be quite far off from that obtained by our method. Figures 6.3 and 6.4, on the other hand, show that for larger values of *n* this difference is greatly reduced. Significantly, for *n* = 1500 (Figure 6.4) the boundary obtained by the proposed classifier is seen to approach the Bayes boundary quite closely.

6.6   Conclusions

A theoretical investigation is made here to find the relationship between a GA-based classifier developed by the authors in References [4] and [5] and the Bayes

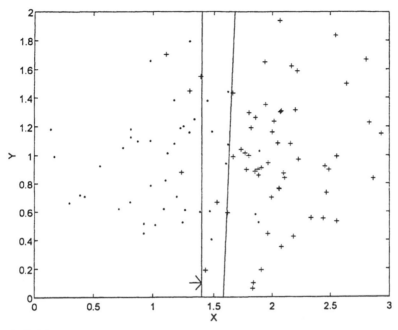

FIGURE 6.2
Bayes decision surface (shown by an arrow) and that obtained by the GA-based classifier for $n = 100$. Class 1 is represented by o and class 2 by +. Both the classes follow the triangular distribution shown in Figure 6.1.

maximum likelihood classifier. It is shown that for a sufficiently large number of sample points and for a sufficiently large number of iterations, the GA-based classifier will try to approach the Bayes classifier.

It is also shown that if we start the algorithm with H_1 hyperplanes, where $H_1 \geq H$, the optimal value, then the scheme for deletion of redundant hyperplanes ensures that we can always arrive at a value H_2 of hyperplanes such that $H_2 = H$, provided there is no confusion in the Bayes decision. Otherwise, H_2 will be greater than or equal to H (see Section 6.4). In either case, the classifier will still be optimal in terms of the number of misclassified sample points.

The experimental results for a triangular distribution of data points show that the decision regions provided by the GA-based classifier do gradually approach the one provided by the Bayes classifier for a large number of sample points. For small values of n, the GA classifier yields a significantly better recognition score. This is due to the reason that such a relatively small data set can be generated by many distributions other than triangular, each providing a different misclassification error function. As the number of points increases, the number of distributions able to generate the points becomes smaller. In the limiting case, the mentioned triangular distribution will be the only one providing the optimal misclassification error. The

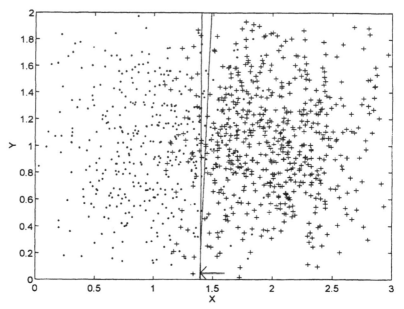

FIGURE 6.3
Bayes decision surface (shown by an arrow) and that obtained by the GA-based classifier for $n = 1000$. Class 1 is represented by o and class 2 by +. Both the classes follow the triangular distribution shown in Figure 6.1.

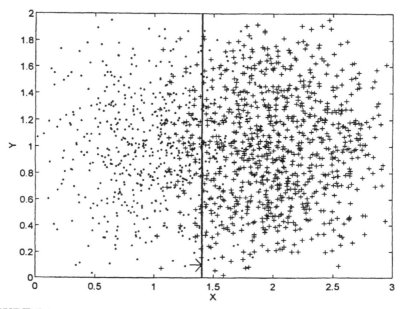

FIGURE 6.4
Bayes decision surface (shown by an arrow) and that obtained by the GA-based classifier for $n = 1500$. Class 1 is represented by o and class 2 by +. Both the classes follow the triangular distribution shown in Figure 6.1.

Table 6.1 Comparative Classwise and Overall Recognition Scores (%)

Total no. of data points	Class	GA-based classifier	Bayes classifier	k-NN classifier (for k = 1)
	1	83.93	76.84	76.72
100	2	94.83	92.38	84.80
	Overall	90.50	86.00	81.50
	1	79.09	79.65	75.55
200	2	95.21	90.65	85.00
	Overall	89.00	86.50	81.50
	1	82.25	81.24	78.66
500	2	92.13	91.32	86.08
	Overall	88.30	87.40	83.20
	1	82.87	81.28	80.43
1000	2	92.90	92.42	87.98
	Overall	88.95	88.00	85.00
	1	82.19	81.37	80.46
1500	2	92.61	92.56	86.23
	Overall	88.40	88.03	83.90

performance of the GA-based classifier, in such a case, will really tend to that of the Bayes classifier. This trend is borne out by the results, where for $n = 1500$, the Bayes regions have almost been obtained. Although only triangular distribution is considered while providing the results, the conclusions will hold for any other distribution.

Acknowledgments

This work was carried out when Ms. Sanghamitra Bandyopadhyay held a fellowship awarded by the Department of Atomic Energy, Government of India, and Prof. S. K. Pal held the Jawaharlal Nehru Fellowship.

References

[1] Goldberg, D.E., *Genetic Algorithms: Search, Optimization and Machine Learning,* Addison-Wesley, New York, 1989.

[2] Davis, L., *Handbook of Genetic Algorithms,* Van Nostrand Reinhold, New York, 1991.

[3] Belew, R.K. and Booker, J.B., Eds., *Proc. 4th Int. Conf. Genetic Algorithms,* University of California, San Diego, Morgan Kaufmann, San Mateo, CA, July 1991.

[4] Bandyopadhyay, S., Murthy, C.A., and Pal, S.K., Pattern classification using genetic algorithms, *Patt. Recog. Lett.,* 16, 801, 1995.

[5] Pal, S.K., Bandyopadhyay, S., and Murthy, C.A., Genetic algorithms for generation of class boundaries, *IEEE Trans. Syst. Man Cybern.,* communicated.

[6] Tou, J.T. and Gonzalez, R.C., *Pattern Recognition Principles,* Addison-Wesley, Reading, MA, 1974.

[7] Bhandari, D., Murthy, C.A., and Pal, S.K., Genetic algorithm with elitist model and its convergence, *Int. J. Patt. Recog. Art. Intell.,* accepted.

[8] Duda, R.O. and Hart, P.E., *Pattern Classification and Scene Analysis,* John Wiley & Sons, New York, 1973.

[9] Apostol, T.M., *Mathematical Analysis,* Narosa Publishing House, New Delhi, 1985.

7

Genetic Algorithms and Recognition Problems

Hugo Van Hove and Alain Verschoren

Abstract   Ordinary linear genetic algorithms do not appear to function optimally in the framework of image processing. In order to remedy this, we introduce and study variants of the genetic algorithm both for trees and two-dimensional bitmaps.

7.1   Introduction

It is well known that the functionality of genetic algorithms (GAs) highly depends on finding a suitable encoding for the data occurring in the optimization problem to be solved. Although advocates of the traditional binary encoding still argue that this way of encoding maximizes the number of schemas sampled by each member of the population, which implies the implicit parallelism of the GA to be used to the fullest, it is becoming more and more apparent that the basic binary encoding is not always appropriate nor practical.

In particular, in the framework of image processing (with typical features like image reconstruction, clustering, and compression), it is obvious that the one-dimensional (linear) genetic algorithm (1DGA) cannot be applied in a natural way, the main obstruction residing in the fact that one loses two-dimensional correlations if one encodes an image linearly. In general, it is well known that forcing data to be encoded in an unnatural way into a one-dimensional string increases the conceptual distance between the search space and its representation, and thus makes it necessary to introduce extra, problem-specific operators (like linear inversion, [4, 5], PMX, [6], or edge recombination, [18], e.g.). Typically, when working with two-dimensional data, if one encodes an image by concatenating its lines, for example, then the classical crossover operator may cause large disruptions vertically. It is thus clear that *no* natural encoding of an image allows us to apply the traditional genetic operators successfully.

On the other hand, when working with data used in pattern or image recognition, one frequently has to deal with *triggers*, i.e., a limited number of features of these data, which are already sufficient to classify the data one considers. As an example, there are $2^{54} \approx 10^{16}$ theoretically possible 6×9 pixel printer characters. So, if we just want to distinguish between a limited number of characters (the ten digits 0 to 9, say), then it is obvious that the corresponding bitmaps contain an enormous amount of redundant information. One should thus be able to determine the character corresponding to a bitmap by checking on a restricted number of well-chosen pixels only. Hence, the corresponding classification problem reduces to constructing a (binary) decision tree, whose leaves are the characters one wishes to recognize and whose nodes reduce to these pixels (the triggers), still to be determined from the data. If one associates a cost to checking the parity of a single pixel (possibly depending on the chosen pixel), then the recognition problem reduces to an optimization problem. This problem may not be solved by applying ordinary GAs, however, essentially because the underlying decision trees do not possess any natural binary encoding, which guarantees the traditional crossover operators always to generate strings, still corresponding to decision trees. This and other reasons force us to work directly with GAs on binary trees.

This text is organized as follows. In the first section, we define and study n-recognition trees and introduce the necessary genetic operators, in order to deal with recognition problems afterward. One of the main features in this setup is a version of the schema theorem, which yields theoretical evidence for the functionality of the algorithm.

In the second section, we briefly study recognition problems and show how the previous machinery may be used to tackle these. The results in this section are somewhat sketchy, we refer to References [14] and [15] for a more detailed treatment.

As pointed out before, when working with images, one benefits by using genetic operators directly on the two-dimensional data, without first encoding these into linear strings. This is particularly true if the tests occurring in a recognition process and used as labels for the vertices in the associated recognition trees are of this type. The micro-operators used in the generation of these (see below) will make use of suitable crossover operators. We have included a short appendix where we included some background concerning these matters.

7.2 Binary Trees

7.2.1

GAs for trees have mainly been used in the context of genetic programming (GP), as in Reference [8] for example, a relatively recent technique, proven to be a rather versatile tool for automatic program generation. Essentially, GP starts from a population of randomly generated LISP programs, represented by their parse trees, and applies genetic operators to make these trees (and the programs they represent) evolve through time. The programs and the corresponding trees are constructed from (1) a set of functions and (2) a set of terminal symbols, defined by the user. The root and the internal nodes of the tree are labeled by functions, with one or more arguments, which may be other functions or terminal symbols. The leaves (or end nodes) of the tree are labeled by terminal symbols and represent input variables, constants, or measurements. Selection of candidates is performed as usually, with a probability of selection proportional to their fitness, which measures the performance of the program with respect to the problem to be solved. Crossover swaps randomly selected subtrees of each chosen parent pair and mutation replaces subtrees in existing members of the population with randomly generated subtrees.

7.2.2

Although GP has mainly been studied "under laboratory conditions," i.e., in situations where the solution is essentially known, Tackett [13] has shown that the same philosophy also functions (with very promising results) when dealing with real life problems (which suffer from noise, poor quality measurements, and unknown answers, and thus include no bias toward optimization of initial parameters, for example). The applications considered in Reference [13] focus on the development of processing trees for the classification of features extracted from images, where input images may be coarse and unreliable, and compare favorably to well-established methods like back-propagation or binary tree methods.

We should point out, as did Tackett [13], that optimal solution trees frequently contain repeated "structures," which are not generated at the outset, but which seem to indicate the existence of structural components, which guarantee (or at least indicate) high performance. On the basis of this, Tackett suggests a version of Holland's schema theorem [7] for trees, a result which we actually include below.

7.2.3

As a last, preliminary example, let us mention recent work by Nguyen [10], who developed a system which permits 3-D models (of planes, for example) to evolve through time, eventually producing models which are more desirable or more fit than the initial ones.

The algorithm starts with crude, randomly generated items, using a model grammar with generic design rules, and applies genetic operators to let these models evolve. The model is based on the following components:

1. *Grammar based, structured object modeling.* Typically, a plane is defined as composed of body, wings, engines, etc., each of these components graphically being represented by geometric primitives. The underlying grammar fully expands the root symbol "plane" into terminal symbols and these are translated into calls to graphical functions.

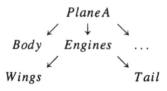

2. *Parametrization of numerical model features and constraints.* These are the numerical parameters in the model (the size of components, e.g.); they are coded into binary strings, in order to allow traditional GAs to be applicable to them.

3. *Performance feedback*, which contributes to shape the path of model evolution. This feedback may either be human or acquired through an automatic recognition process.

The evolution engine in the model acts on two levels:

1. *At the structural level.* The model design is expressed as a tree, without any direct numerical values. Crossover is performed between subtrees of the same functionality (e.g., only between subtrees with top "engines").

2. *At the microlevel.* In order to change the numerical value of the parameters (expressed as a binary string), crossover is applied to strings, which refer to the same parameter (e.g., the length of the "tail" component).

7.2.4

The methods we are about to develop are somewhat different from the previous ones.

1. Just as in Reference [13], we restrict to binary trees. Of course, even in GP, this is no real restriction, as this essentially just amounts to consider

only the primitive (car, cdr)-pairs, within the linked list structure one uses to encode more complex lists. It appears that this restriction not only makes implementations easier, but also increases performance (cf. binary vs. n-ary string encoding for traditional GAs).

2. We only consider fixed width trees. This restriction is rather serious, but necessary, in order to avoid redundancy problems when tackling the recognition problems we will describe below. On the other hand, even in GP and program design, where it seems unreasonable to work with fixed (or limited) width trees only, there is a parsimony argument cf. Reference [13], which suggests adding width as a component to the fitness function, not only for "aesthetic" or simplicity-of-solution reasons, but also for its intrinsic relationship to fitness, i.e., in view of some internal bound on the "appropriate size" of the solution tree. This is also the point of view adopted in Reference [11], to which we refer for a treatment of variable width trees, in the framework of game theory and optimal strategy development.

3. We have to introduce extra operators, in order to handle questions arising from image and pattern recognition. Indeed, whereas, for example, the mutation operator used in GP provides enough *external* dynamics (by importing new random program chunks), we need to include more *internal* dynamics, which slightly change reasonable subprograms or move these to better locations within the program. This leads us to introduce operators like "switch" (which inverts the orders of the arguments of a function) or "translocation" (which moves subtrees to a different location or allows to use a different root, without losing well-performing schemata).

7.2.5

Let us first formally describe the trees on which the GA will act. We refer to References [14] and [15] for more details and implementation aspects. Fix a positive integer n and an arbitrary set P. (In practice, P will usually be the set $\{0, 1, \Box\}^\ell$ of length ℓ strings or $\{0, 1, \Box\}^{r \times s}$ of the set $(r \times s)$-bitmaps, both with "don't care character" $\Box$, and its elements should be viewed as tests to be performed during a classification or recognition process, for example.) We formally adjoin to P a terminal symbol $*$ and denote the resulting set $P \cup \{*\}$ by $\widehat{P}$.

A *recognition tree* of width n (or simply an *n-recognition tree*) over P is a quadruple $A = (V, E, \varepsilon, \delta)$, consisting of:

1. A binary tree (V, E) with edge set E, vertex set V, and with a cardinality n set $T \subseteq V$ of terminal nodes.

2. An injective map
$$\varepsilon : (V - T) \times \{+, -\} \to E.$$

For any $v \in V - T$, we call $\varepsilon(v, +)$ resp. $\varepsilon(v, -)$ the *left* resp. *right* edge starting at v.

3. A map $\delta : V \to \widehat{P}$, with $\delta^{-1}(*) = T$. In other words, every nonterminal node is labeled by an element of P, whereas the terminal nodes are labeled by $*$.

One may graphically represent a recognition tree by applying δ and ε to label the vertices and edges of the underlying tree (V, E). For example:

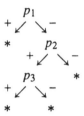

7.2.6

Let us start from a population $\mathcal{A}$ of n-recognition trees for some fixed positive integer n and assume we are given a fitness function f, which to any recognition $A \in \mathcal{A}$ associates its fitness $f(A) \in \mathbf{R}_+$. As we will see below, usually f contains a global component, which evaluates the overall shape of the tree, and a local component, which relates to the fitness of elements in the tree.

For any $v \in V$, denote by A_v the subtree of A with "top" v and by $n_A(v)$ its width. The following *global* genetic operators on n-recognition trees may then be introduced:

1. The *reproduction operator* selects recognition trees in $\mathcal{A}$ with a probability proportional to their fitness and is thus completely similar to the usual reproduction operator. In particular, any $A \in \mathcal{A}$ has a probability

$$p(A) = \frac{f(A)}{\sum_{B \in \mathcal{A}} f(B)}$$

of being selected.

2. The *crossover operator* is the recognition tree analogue of the crossover operator in the classical GA.
    Starting from recognition trees $A = (V, E, \varepsilon, \delta)$ and $A' = (V', E', \varepsilon', \delta')$, one randomly selects an edge $e : w \to v$ in E, where $v \notin T$, the terminal nodes of A. If $n_{A'}(v') \neq n_A(v)$ for all $v' \in V'$, crossover does nothing. Otherwise, select v' arbitrarily amongst the vertices $v' \in E'$ with $n_{A'}(v') =$

$n_A(v)$ and replace A and A' by the n-recognition trees $B = (W, F, \mu, v)$ resp. $B' = (W', F', \mu', v')$ obtained by "replacing" $e : w \rightarrow v$ by a new edge $f : w \rightarrow v'$ and A_v by A'_v, and similarly for B'. If e is the right resp. left edge starting from w in A, then so will be f in the new tree B.

Example: The 4-recognition trees with p_2 and q_2 as crossover vertices:

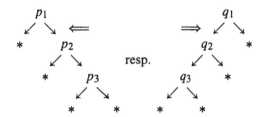

yield through crossover the following 4-recognition trees:

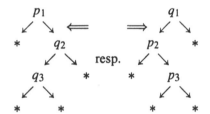

An important variant of the previous operator, which does not cause any "grammatical" disruption, only permits to exchange subtrees A_v and $A'_{v'}$ if $n_A(v) = n_{A'}(v')$ and $\delta(v) = \delta'(v')$.

3. The *switch operator* on recognition trees is a global analogue of the mutation operator on strings. Mutation is a (low probability) operator which occasionally switches a bit from 0 to 1 or from 1 to 0 and thus improves the dynamics of the genetic algorithm.

    The switch operator depends on the choice of a vertex v in an n-recognition tree $A = (V, E, \varepsilon, \delta)$. If $v \in T$, it does nothing. Otherwise, it replaces A by the n-recognition tree

$$A^s_v = (V, E, \varepsilon^s_v, \delta),$$

where $\varepsilon^s_v : (V - T) \times \{+, -\} \rightarrow E$ coincides with ε, except in v, where $\varepsilon^s_v(v, +) = \varepsilon(v, -)$ resp. $\varepsilon^s_v(v, -) = \varepsilon(v, +)$.

Example: The switch operator applied to the node p_2 replaces the recognition tree:

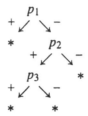

by the following recognition tree:

4. The *translocation operator* on recognition trees may be viewed as an auto-crossover operator and is a tree analogue of linear inversion [4] as well as of translocation [12] in the ordinary GA on strings.

Translocation starts from an n-recognition tree $A = (V, E, \varepsilon, \delta)$ and some $v \in V$ resp. $t \in T$, and replaces A by a new recognition tree:

$$A_{v,t} = (V_{v,t} = V, E_{v,t}, \varepsilon_{v,t}, \delta_{v,t} = \delta).$$

The edge set $E_{v,t}$ consists of the edges $e_{v,t}$, one for each $e : v_1 \to v_2$ in E, defined as follows: (1) if $v_2 \neq v, t$, then $e_{v,t} = e$; (2) if $v_2 = v$, then construct a new edge $e_{v,t} : v_1 \to t$; (3) if $v_2 = t$, then construct a new edge $e_{v,t} : v_1 \to w$, with $w = \top$ (the top of A) or $w = v$, depending on whether $t \in A_v$ or not.

The map $\varepsilon_{v,t}$ is defined in the obvious way, i.e., by putting $\varepsilon_{v,t}(w, +) = \varepsilon(w, +)_{v,t}$ resp. $\varepsilon_{v,t}(w, -) = \varepsilon(w, -)_{v,t}$, for any $w \in V - T$.

Example: If we apply translocation at the indicated nodes in the following recognition tree:

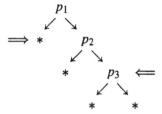

then we obtain the following recognition tree:

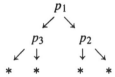

On the other hand, starting from the selected nodes:

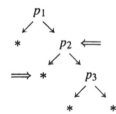

we obtain the following recognition tree (with different[1] top-node!):

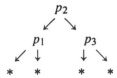

---

[1]The translocation operator we just defined differs slightly from the one previously used in Reference [15]. Indeed, whereas the present operator permits a change of top-node, the one in Reference [15] did not. This extra flexibility allows for more dynamics and faster convergence of the associated GA.

7.2.7

The previous operators only modify the overall structure of recognition trees, but do not affect the contents of their nodes. In order to remedy this, we have to introduce some "noise", i.e., some low probability micro-operators which directly act on the nonterminal values of δ. As these *local* operators may cause big disruptions, due to the fact that they can completely change the "meaning" of decision trees used in recognition problems, for example (since they internally modify the tests corresponding to yes/no nodes and thus lead to a complete reinterpretation of well-shaped trees), they should be handled with extreme care. Usually, micro-operators are introduced in the beginning of the genetic process, in order to guarantee enough internal dynamics and, just as in processes regulated by simulated annealing, their influence should gradually be constrained at later stages.

Typically, their probability of application should be regulated by a time- or fitness-dependent factor, like $P_\alpha = 1 - (\pi/2)^{-1} Arctg(\alpha t)$ or $P_\alpha = e^{-\alpha t}$, where the relaxation parameter α may be user or process defined.

Although other micro-operators yield rather satisfactory results as well, in our implementations, we have essentially restricted to the following basic operators:

1. **Micro-crossover** depends upon a map

$$\chi : P \times P \times J \to P \times P,$$

   where $J = [0, 1]$. In many applications, P will be the set $\{0, 1, \square\}^\ell$ or $\{0, 1, \square\}^{r \times s}$, and for any $j \in J$, the induced map $P \times P \to P \times P$ will be a crossover operator of a type or at a locus determined by j (see also the appendix on 2DGAs).
   Starting from a couple of n-recognition trees $A = (V, E, \varepsilon, \delta)$ and $A' = (V', E', \varepsilon', \delta')$, one selects nonterminal vertices $v \in V$ and $v' \in V'$. If $rand \in J$ is a random variable, then micro-crossover replaces $\delta(v)$ resp. $\delta'(v')$ by q resp. q', determined by $(q, q') = \chi[\delta(v), \delta'(v'), rand]$. In its coarse version, v and v' are selected randomly. A more refined version of the operator considers P as consisting of *(type, value)*-couples, and only allows micro-crossover between pairs with the same *type* component.

2. **Micro-mutation** is determined by a map

$$\mu : P \times J \to J,$$

   with J as before. If $P = \{0, 1, \square\}^\ell$, one may view μ as changing the bit at a location determined by j, through the induced map $P \to P$. In the general case, one replaces for any $v \in V - T$, the value $\delta(v)$ by $\mu[\delta(v), rand]$ (where $rand \in J$ is again a random variable).

7.2.8

The complete GA for n-recognition trees functions as the traditional GA, i.e., starting from a random population $\mathcal{A}(0)$ of n-recognition trees, one constructs the population $\mathcal{A}(t + 1)$ from the previous population $\mathcal{A}(t)$, by selecting through reproduction $A, A' \in \mathcal{A}(t)$ and applying successively

1. Crossover, with probability p_c

2. Switch, with probability p_s

3. Translocation, with probability p_t

4. Micro-operators, with probabilities $p_{\mu c}$ and $p_{\mu m}$

Of course, there are several variants of this general description which should be fine-tuned, depending on the application one considers (cf. Reference [14]). In particular, one needs to decide whether switch and translocation should be applied to a single node or to all nodes (with suitable probabilities) and whether the coarse or a more refined version of the micro-operators should be used.

7.2.9 The Schema Theorem

The performance of the classical GA on length ℓ strings may best be described by using schemata, i.e., elements of $\{0, 1, \square\}^\ell$.

One says that a string $A \in \{0, 1\}^\ell$ satisfies the schema H (and one writes this as $H \rightarrow A$), if A coincides with H at all places different from $\square$. The quality of a schema H, i.e., of the strings satisfying H, with respect to the solution we aim at, may then be measured by the average fitness $f(H, t)$ of H in the t th generation, where

$$f(H, t) = \sum_{H \rightarrow A} f(A)/m(H, t),$$

with $m(H, t)$ the number of strings in $\mathcal{A}(t)$ satisfying H. The *schema theorem* [7] proves that above-average schemata will essentially receive an exponentially increasing number of representatives throughout the successive generations.

In order to generalize this setup, let us first introduce the notion of embedding between recognition trees (cf. Reference [15]). So, consider recognition trees $A = (V, E, \varepsilon, \delta)$ and $H = (V_H, E_H, \varepsilon_H, \delta_H)$ over the same set P and denote by T resp. T_H the terminal nodes of A resp. H.

An *embedding* $H \rightarrow A$ is a pair of maps $\alpha : V_H \rightarrow V$ resp. $\beta : E_H \rightarrow E$ satisfying:

1. For all $e : v_1 \rightarrow v_2$ in E_H, we have $\beta(e) : \alpha(v_1) \rightarrow \alpha(v_2)$ in E; in particular, $\alpha(V_H - T_H) \subseteq V - T$.

2. The maps α and β fit into a commutative diagram:

$$(V_H - T_H) \times \{+, -\} \xrightarrow{\alpha \times id} (V - T) \times \{+, -\}$$

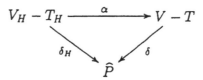

3. The map α makes the following diagram commutative:

Let us now define an *n-recognition schema* to be just an *r*-recognition tree, with $r \leq n$. If H is an n-recognition schema, then any n-recognition tree A endowed with an embedding $H \to A$ is said to satisfy or to represent H. Intuitively, this just says that A "contains a copy" of H.

The philosophy behind this is that the occurrence of repeated "patterns" within high fitness recognition trees might be the basic reason of optimal behavior, just as claimed in a similar setup in Reference [13].

To make this more concrete, consider an n-recognition schema H, and denote by $A(H, t)$ the set of all representatives of H in the nth generation $A(t)$, obtained by repeatedly applying the GA to an initial random population $A(0)$ of n-recognition trees. We want to describe the behavior of $m(H, t) = |A(H, t)|$ through the genetic process.

Actually, if the genetic operators are applied as described in the above algorithm, with their associated probabilities, then one may prove using some combinatorics to derive probabilities of destruction and survival of schemata (cf. Reference [15]) that $m(H, t + 1)$ is at least equal to

$$m(H, t) \frac{f(H)}{\bar{f}} (1 - p_c \frac{w - 2}{n - 2})(1 - p_s \frac{w - 1}{n - 1})(1 - p_t \frac{w - 2}{n - 2})(1 - p_\mu \frac{w - 1}{n - 1}),$$

where $w = w(H)$ is the width of H; see Reference [15] for details. If we put

$$p_M = p_s + p_{\mu c} + (n - 1)p_{\mu m} = p_s + p_\mu,$$

resp.

$$p_C = p_c + p_t$$

and if we ignore small cross-product terms, then we find

$$m(H, t+1) \geq m(H, t)\frac{f(H)}{\overline{f}}(1 - p_C\frac{w-2}{n-2} - p_M\frac{w-1}{n-1}).$$

In particular, assuming (as a first approximation) $f(H)/\overline{f}$ to remain constant for a number of generations, this proves the following version of the schema theorem: *n-recognition schemata of low width and above average fitness ("building blocks") will receive an exponentially increasing number of trials in subsequent generations.*

7.3 Recognition Problems

7.3.1

Although it seems intuitively clear what we mean by *recognizing* and *classifying* objects, modern psychology shows us that it is actually very hard to concretely and formally define these notions.

Essentially, the problem consists in considering a class of objects and of defining a partition in this class, corresponding to certain properties which these objects are supposed to satisfy. If we wish to make a classification based on a single property, this partition will consist of two classes, where one class contains the objects which satisfy the property and the other class the objects which do not. Conversely, the *property* or *characteristic* we wish to investigate could formally simply consist of checking whether an object does or does not belong to one of the classes of a given partition. Technically, properties defined in a given class of objects are thus just partitions of this class.

However, in concrete situations, or at least in situations where the classification problem may be automated, recognizing or classifying objects amounts to *measuring* certain well-defined features, which the objects we wish to study do or do not possess. In other words, for each object, we are given a tuple of numerical data and the classification of these objects is realized by fixing threshold values for certain components of this tuple. (For example, animals or animal types may be described by arrays containing their number of legs, wings, the color of their skin, number of teeth, presence or length of a tail. Clearly, all of these data may be quantified.)

In what follows, objects will be identified with one- or two-dimensional arrays, which quantitatively describe their salient features, obtained, for example, by measuring these properties by some external means.

Of course, this measuring has a price, be it a real cost or simply the time necessary to perform the action of measuring, the amount of personnel, or material needed for it. It is thus extremely important to minimize the number of data needed to realize the classification or recognition of the test objects presented to the user. In particular, it is necessary to investigate whether some of the measured features in the test data are linked or, even worse, are redundant for the classification problem. It is indeed clear (as pointed out in the introduction) that if we work with a small collection of test objects, each of these objects being described by a large number of features, then in practice a limited number only of features of these objects will have to be studied in order to separate them.

7.3.2

Let us start from a finite set S of objects, endowed with a family of maps $\mathcal{F} = \{f_i : S \to D_i;\ i \in I\}$. The maps f_i are called *features* and should be viewed as measuring properties of the objects in S. In particular, one will usually take for D_i one of the sets $\{0, 1\}$, $[0, 1]$, or $\mathbf{R}$ (the real numbers). We will also assume the feature set $\mathcal{F}$ to be endowed with an $\mathbf{R}$-valued *expense function* E, which to any feature f_i associates the cost of evaluating it on some object[2] in S.

A subset $T \subseteq S$ is said to be a *test space* if the map

$$F : T \to \prod_{i \in I} D_i : t \mapsto [f_i(t);\ i \in I]$$

is injective. In other words, the objects in T should completely be determined by their features.

As an example, let S be the set of all $(r \times s)$-bitmaps and let $f_i : S \to \{0, 1\}$ be a set of parity testers, i.e., each f_i just checks the value of the bit at a fixed location, which depends on i. A test space $T \subseteq S$ is then just a collection of bitmaps, which may be distinguished by using the feature set $\mathcal{F}$ (a specific collection of bits).

7.3.3

A *test* (over I) is a couple $\tau = (i, \beta)$, with $i \in I$ and $\beta \in \overline{D_i}$, with $\overline{D_i} = D_i \cup \neg D_i$, where $\neg D_i$ consists of symbols $\neg\alpha$, one for each $\alpha \in D_i$. A *pattern* (over I) is a (finite) family $\pi = \{\tau_1, \dots, \tau_r\}$ of tests.

We say that $s \in S$ satisfies the test $\tau = (i, \beta)$ if either $\beta \in D_i$ and $f_i(s) = \beta$ or $\beta \in \neg D_i$, i.e., $\beta = \neg\alpha$ for some α and $f_i(s) \neq \alpha$. We say that $s \in S$ satisfies the pattern π as above, if s satisfies every test τ_i in it.

For example, if we again consider the set S of $(r \times s)$-bitmaps with features corresponding to checking particular bits, then a pattern may obviously be represented by an $r \times s$-matrix with entries in $\{0, 1, \square\}$, where the entries different from $\square$ correspond to the tests (and their outcome-values) in π. In this way, an element

---

[2]For simplicity's sake, we assume E to be homogeneous, in the sense that $E(f_i)$ is independent of the object in S on which f_i is evaluated.

$s \in S$ will satisfy the pattern π, if it only differs from the corresponding matrix in entries where the latter has $\square$.

The expense function $E : \mathcal{F} \to \mathbf{R}$ may be extended to the set of all tests resp. patterns, by putting $E(\tau) = E(f_i)$, if $\tau = (i, \beta)$ resp. $E(\pi) = \sum_{i=1}^{r} E(\tau_i)$, with $\pi = \{\tau_1, \ldots, \tau_r\}$ as before.

7.3.4

Let us now consider an n-recognition tree $A = (V, E, \varepsilon, \delta)$ over P, where $n = |T|$, the cardinality of a test space $T \subseteq S$, and where P is the set of all patterns over I. We say that $t \in T$ is (uniquely) recognized by A within T, if there is a unique path

$$\mathsf{T} = v_0 \to v_1 \to \ldots \to v_r \to *,$$

such that t satisfies all patterns $\delta(v_i) \in P$ and if t is the only element in T with this property. If A recognizes all elements in T, then we say that A *recognizes* T. In particular, A is then just a decision tree for T, permitting to distinguish the elements in T (which correspond to the end nodes of A), by checking the patterns which label its vertices.

It is clear from our assumptions that for any test space $T \subseteq S$, one may produce some n-recognition tree, which recognizes T. On the other hand, we would like to find a solution to the *recognition problem* (for T), i.e., we want to construct a *minimal* n-recognition tree A for T, in the sense that it not only recognizes T, but that, moreover, its "cost" $E(A) = \sum_{path} E(path)$ is minimal, where for any path

$$\mathsf{T} = v_0 \to v_1 \to \ldots \to v_r \to *,$$

we put $E(path) = \sum_{i=0}^{r} E[\delta(v_i)]$.

If the expense function E is constant, then $E(A)$ only depends on the length of the paths in A and it is not hard to find a rather straightforward algorithm to solve the recognition problem (cf. Reference [15]). On the other hand, if E is not constant, let us define a fitness function f on n-recognition trees over P by putting

$$f(A) = \alpha E(A) + \beta[n - n(A)],$$

where $n(A)$ is the number of elements in T, which are recognized by the n-recognition tree A, where $E(A)$ is as before, and where α and β are user-defined parameters, which allow to stress the importance of cost vs. number of correctly recognized items. Starting from a randomly generated population of n-recognition trees over P, it should then be clear that the genetic algorithm described in the previous section thus allows for an efficient solution in this situation as well. For other applications, we refer to Reference [14].

7.4   Appendix: Two-Dimensional Genetic Algorithms

A.1

It has already been mentioned in the introduction that we clearly benefit by using two-dimensional algorithms (2DGAs) when we work with data stemming from image processing, the reason for this being that linear encoding of two-dimensional material eliminates nonlinear correlations.

Similar considerations urged Cartwright and Harris [1] to work in a two-dimensional setup, in order to solve the so-called *source apportionment problem* by using GAs. Let us briefly describe their ideas.

In the source apportionment problem, one considers a number of sources of pollution in some fixed area, which all contribute to the overall pollution in a varying degree. Receptor stations, situated at different locations, are monitoring the situation by taking samples regularly. The question is to determine the quantity of pollution caused by the individual pollutants from an analysis of these samples. (Another model for the same problem consists of monitoring car traffic between different cities, in order to determine the origin and destination of travelers.) This appears to be a very complex problem. Actually, even the best traditional methods still yield deviations up to 30%. The GA method developed in Reference [1] typically yields quantitative results with a deviation of about 0.3%.

If there are r receptors and s sources, then a solution of the problem may be represented by an $(r \times s)$-matrix

$$P = \begin{pmatrix} p_{11} & \cdots & p_{r1} \\ \vdots & \ddots & \vdots \\ p_{1s} & \cdots & p_{rs} \end{pmatrix},$$

where p_{ij} represents the fraction of the total pollution, emitted by source j and sampled at receptor i. This solution (or an approximation) is found by applying genetic operators to a random initial population of matrices, where the fitness of a matrix in one of the successive generations depends on its deviation with respect to the actual solution.

Encoding P as a string

$$(p_{11} \cdots p_{r1} p_{12} \cdots \cdots p_{rs})$$

and applying the ordinary 1DGA does not work. Indeed, whereas the pollutant profile of a single source corresponds to a scheme

$$(\square \ldots \square p_{1j} \ldots p_{rj} \square \ldots \square),$$

with contiguous alleles (and hence is reasonably resistant to crossover disruption), the receptor profile of a single source corresponds to a scheme

$$(\Box \ldots \Box p_{i1} \Box \ldots \Box p_{i2} \Box \ldots \Box p_{is} \Box \ldots \Box),$$

with noncontiguous alleles (and hence is highly *non*resistant to crossover disruption).

Now, as was already pointed out in Reference [1], even when working directly with two-dimensional data, one still has to be careful about the choice of a suitable crossover operator. The naive way of generalizing the traditional, linear crossover operator would be to choose two crossover sites (x_1, y_1) resp. (x_2, y_2) and to interchange the block determined by these end points with the corresponding one of another element of the current population. However, since Cartwright and Harris apply a selective GA (only keeping one child after each crossover), just as one uses wraparound crossover in the one-dimensional case, in order to uniform sampling, they apply the so-called "UNBLOX" crossover, which, depending on whether $x_1 < x_2$ or $x_1 > x_2$ resp. $y_1 < y_2$ resp. $y_1 > y_2$, interchanges the blocks indicated in the figure below. Sampling with respect to the latter operator is uniform and yields the highly successful results mentioned above.

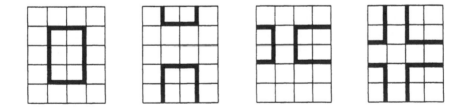

A.2

Two-dimensional genetic operators have also been applied successfully by Cherkauer [3], in order to obtain good examples for a nearest-neighbor pattern classifier. The *Genncon*[3] algorithm and its particular implementation *GC1* classifies two-dimensional rectangular bitmaps, using domain knowledge to initialize and to guide the search for solutions. In order to calculate the similarities between a "rule" and a "training example," both of the bitmaps are of the same size, one in principle just counts the number of matching pixels. To make this metric "shift-resistant," a somewhat different approach is needed. In the implementation,

---

[3]Genetic Nearest-Neighbor exemplar CONstruction.

Cherkauer thus allows one-pixel shifts in each of the eight primary and secondary compass directions and uses an ingenious similarity metric to evaluate matching.

Starting from an initial "rule base" (initialized with a subset of training examples), to speed up convergence and to increase efficiency, *GCl* uses mutation and crossover to obtain a higher NN classification accuracy. Mutation just flips pixels randomly, whereas crossover interchanges the pixels inside corresponding rectangular regions in parent pairs

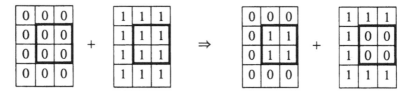

On the other hand, unlike classical GAs who use some form of reproduction, depending on the value of a fitness function, *GCl* is applied uniformly and thus encourages diversity. As in the previous example, much of the strength of the algorithm resides in an adequate choice of crossover operators.

A.3

The previous crossover operators may also be applied when dealing with images, i.e., $(r \times s)$-bitmaps, at least when we use a selective GA (only keeping one child of each parent pair) or a uniform GA (without reproduction). If we want to apply an analogue of the classical GA, we need somewhat different operators, which are two-dimensional versions of one-point crossover for strings and which permit to treat large chunks of an image as well as smaller details. In our implementation, the following three variants of the classical crossover operator were used:

1. The crossover sites for *straight crossover* will be at the end of each row and at the bottom of each column. It chooses one of these sites randomly, i.e., each of these $(r-1)+(s-1)$ sites has the same probability of being chosen. If we choose location 2 as crossover site, in the picture below, then the original and resulting bitmaps will be:

The bitmaps are thus cut along a straight line and the resulting components are combined into two new bitmaps.

2. The crossover sites for *crooked crossover* are the $2(rs - 1)$ locations between the bits of the bitmap. In a picture:

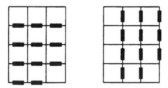

Picking the sixth horizontal separator as crossover site, for example, bitmaps will be modified in the following way:

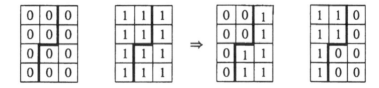

3. *Corner crossover* picks two crossover points, one on the vertical and one on the horizontal edges of the bitmap, and modifies the bitmaps as indicated in the example below:

0	0	0
0	0	0
0	0	0
0	0	0

1	1	1
1	1	1
1	1	1
1	1	1

$\Rightarrow$

1	1	0
1	1	0
0	0	0
0	0	0

0	0	1
0	0	1
1	1	1
1	1	1

A.4

The resulting 2DGA has theoretically and experimentally been compared with the 1DGA applied on the strings obtained by concatenating the rows of the matrices on which it acts (cf. References [14] and [16]). It appears that the 1DGA causes a much bigger disruption in the vertical dimension than the 2DGA. This may also theoretically be seen by calculating the probability that two neighboring bits in the same column are separated through crossover. For the 1DGA this probability is $p_1 = \frac{r}{(rs-1)}$ and for crooked crossover, e.g., this is $p_2 = \frac{1+r}{2(rs-1)}$. This also explains why the convergence rate of the 2DGA is higher than that of the 1DGA for bitmaps.

A.5 The Schema Theorem

Let us fix a two-dimensional schema H, i.e., an $(r \times s)$-matrix with entries in $\{0, 1, \square\}$. The two-dimensional analogue of the notion of *defining length* of H, (cf. References [5, 7, et al.]), will be the dimension (r_H, s_H) of the *defining rectangle* of H, i.e., the smallest rectangle in H, which contains all bits different from $\square$.

Let $m(H, t)$ be the cardinality of the set of matrices representing H in the population at time t and, for any given fitness function f, let us denote by $f(H)$ the average fitness of the matrices in this set. Finally, we write $\overline{f}$ for the average fitness of the matrices in the population at time t.

One may then prove the following version of the *schema theorem* for the 2DGA, (cf. References [14] and [16]):

$$m(H, t+1) \geq m(H, t)\frac{f(H)}{\overline{f}}[1 - p(H) - o(H)p_m],$$

where $o(H)$ is the order of H, i.e., the number of bits in H different from $\square$, where p_m is the mutation probability for individual bits and $p(H)$ the probability that H will not survive crossover. Only this last factor remains to be calculated.

If we assume that the schema H does not survive, if its defining rectangle is touched upon through crossover, it is fairly easy to verify the following values for the probability $p(H)$, where p_c is the probability that crossover occurs:

- Straight crossover:

$$p = p_c\frac{r_H + s_H - 2}{r + s - 2}$$

- Crooked crossover:

$$p = p_c\frac{(r+1)(s_H-1) + (s+1)(r_H-1)}{2(rs-1)}$$

- Corner crossover:

$$p = p_c \frac{(r_H - 1)(s_H - 1)}{(r - 1)(s - 1)}$$

We leave it as a straightforward (combinatorial) exercise to the reader to give a more detailed version of the previous result, if all three crossover operators occur with fixed probabilities.

References

[1] Cartwright, H.M. and Harris, S.P., The application of the genetic algorithm to two-dimensional strings: the source apportionment problem, in Int. Conf. on Genetic Algorithms, Urbana-Champaign, IL, 1993.

[2] Cavicchio, D.J., Adaptive Search Using Simulated Evolution, Doctoral dissertation, University of Michigan, Ann Arbor, 1970.

[3] Cherkauer, K.J., Genetic search for nearest-neighbor exemplars, in Proc. 4th Midwest Artificial Intelligence and Cognitive Science Society Conf., Utica, IL, 1992, 87.

[4] Frantz, D.R., Non-Linearities in Genetic Adaptive Search, Doctoral dissertation, University of Michigan, Dissertation Abstracts International, 33(11), 5240B, University Microfilms No. 73-11 (116), 1972.

[5] Goldberg, D.E., *Genetic Algorithms in Search, Optimization and Machine Learning,* Addison-Wesley, Reading, MA, 1989.

[6] Goldberg, D.E. and Lingle, R., Alleles, loci and the traveling salesman problem, in *Proc. 1st Int. Conf. on Genetic Algorithms and Their Applications,* ICGA-85, Grefenstette, J.J., Ed., Lawrence Erlbaum Associates, Hillsdale, 1985.

[7] Holland, J.H., *Adaptation in Natural and Artificial Systems,* University of Michigan Press, Ann Arbor, 1975.

[8] Koza, J.R., *Genetic Programming,* MIT Press, Cambridge, 1992.

[9] Liepins, G.E. and Vose, M.D., Representational issues in genetic optimization, *J. Exp. Theor. Artif. Intell.,* 2, 101, 1990.

[10] Nguyen, T.C., Goldberg, D.E., and Huang, T.S., Evolvable Modeling: Structural Adaptation through Hierarchical Evolution for 3-D Model-based Vision, Urbana and Beckman Institute and Coordinated Science Laboratory, University of Illinois, preprint.

[11] Robeys, K., Van Hove, H., and Verschoren, A., Genetic algorithms and trees, II. Strategy trees (the variable width case), *Comput. Artif. Intell.,* to appear.

[12] Smith, S.F., A Learning System Based on Genetic Adaptive Algorithms, Ph.D. thesis, University of Pittsburgh, Pittsburgh, PA, 1980.

[13] Tackett, W.A., Genetic Programming for Feature Discovery and Image Discrimination, University of Southern California, preprint.

[14] Van Hove, H., Representation issues in Genetic Algorithms, Ph.D. thesis, University of Antwerp, RUCA, 1995.

[15] Van Hove, H. and Verschoren, A., Genetic algorithms and trees. I. Recognition trees (the fixed width case), *Comp. Artif. Intell.,* 13, 453–476, 1994.

[16] Van Hove, H. and Verschoren, A., Two-Dimensional Genetic Algorithms, to appear.

[17] Vose, M.D., Generalizing the notion of schema in genetic algorithms, *Artif. Intell.,* 50, 385, 1991.

[18] Whitley, D., Starweather, T., and Fuquay, D., Scheduling problems and traveling salesman: the genetic edge recombination operator, in *Proc. 3rd Int. Conf. on Genetic Algorithms,* Schaffer, J.D., Ed., ICG A-89, Morgan Kaufmann, San Mateo, CA, 1989.

8

Mesoscale Feature Labeling from Satellite Images

Bill P. Buckles, Frederick E. Petry, Devaraya Prabhu, and Matthew Lybanon

Abstract   A general approach to the application of genetic algorithms to automatic scene labeling task is presented. The remote sensing application discussed involves labeling the oceanic features of the North Atlantic Gulf into known classes from presegmented images. Heuristic knowledge about the domain constraints is represented by a semantic net, in which nodes represent classes and each directed link represents a predicate from one class to another. Every segment in the image is an instance of some class and inherits all the predicates associated with it. Each solution evolved by the genetic algorithms represents a complete labeling of the image. Fitness of such a solution is computed as a bottom-up measure of the satisfaction of domain constraints.

8.1   Background

Genetic algorithms (GAs) have been applied to several aspects of image understanding. Ignoring the considerable work concerning GAs and neural nets, image applications of GAs may be divided into two categories—those that determine settings for real-valued parameters and those that solve for combinatorial aspects of the problem. Among the former are those that solve problems generally inherent in image understanding, such as enhancement [9], and those that solve for domain-specific parameters. An example is determining heart ventricle orientation [5].

Just as numerous are those applications that exploit combinatorial aspects of pattern recognition. These tend to more frequently address general issues. Perhaps the earliest such application was feature selection [11]. More recently, methods for region-based segmentation based on a state-space search has appeared [10], as well as organizing the objects in a scene [12]. Domain-specific combinatorial problems are represented, however. For instance, determining the subset of antigens that account for a particular configuration of antibodies [4].

The application studied here falls into the combinatorial category. Although presented here in the context of a specific domain—oceanic feature labeling—we are applying the same method to other domains. The labeling of oceanic features has been the subject of other solution methods including constraint relaxation [7, 8]. The research concentrates on satellite images of North Atlantic Gulf Stream region and the infrared band (10.3 to 11.3 μm). Images are preprocessed via edge segmentation using methods discussed in Reference [6]. The segmented image is then labeled, segment by segment, using a known set of classes. An infrared satellite image and its companion segmented image is shown in Figures 8.1 and 8.2. The possible classes for this specific problem are "Gulf Stream North Wall"(N), "Gulf Stream South Wall"(S), "Warm Eddy"(W), "Cold Eddy"(C), and "Other"(O). Domain knowledge necessary for this task is mainly heuristic in nature, as in many other object recognition tasks, and is based on spatial properties of the segments and their relationships with one another, within broad tolerance levels. Representation of this domain knowledge about relationships and methods for quantifying them in a specific context are prerequisites for applying GAs to this task.

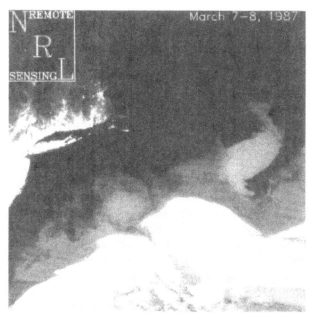

FIGURE 8.1
Original infrared image.

FIGURE 8.2
Segmented version of Figure 8.1.

8.2 Methodology

8.2.1 Problem Model

Binary relationships between concepts are easily modeled by a semantic net (or, more precisely, an association list). Each node of the semantic net represents a class (i.e., label) and each directed arc denotes a relationship from one class to another. Each segment present in the image is an instantiation of one of these possible classes and therefore inherits all the relationships associated with that class. Figure 8.3 illustrates this idea for an arbitrary labeling of the segments shown in Figure 8.2.

The basic scheme is based upon a semantic net which we term a *fitness net* which is illustrated in Figure 8.3. The fitness net is a general model for defining fitness values for a labeling algorithm and is formally defined as:

$$\text{FN} = \langle \text{S, M, SN} = \langle \text{C}, \mathcal{P}, \textbf{null} \rangle, \text{ISA}, \mathcal{F}, \textbf{W}, \textbf{E} \rangle$$

where

S	=	$\{S_1, S_2, \ldots, S_n\}$;　a set of image segments
M	=	$\{m_1, m_2, \ldots, m_h\}$;　a set of features measurable over each segment
SN	:	is a semantic net for which $\mathbf{C} = \{C_1, C_2, \ldots, C_m\}$;　a set of class labels　$\mathcal{P} = \{P_{ijk} \mid 1 \leq i, j \leq m, 1 \leq k \leq h\}$;　a set of predicates, one for each feature and ordered pair of classes　**null** $\in \mathcal{P}$ and **null** $(x) = 1$ for all x
ISA	:	$\mathbf{S} \to \{1, 2, \ldots, m\}$;　an instance function that maps a segment to the index of a class
$\mathcal{F}$	=	$\{f_{ijk} \mid 1 < i, j < m, 1 \leq k \leq h\}$;　a set of feature comparison functions, one for each predicate
W	=	$\{w_{ijk} \mid 1 < i, j < m, 1 \leq k \leq h\}$;　a set of weights, one for each predicate
E	:	an evaluation function

Because the relationships (i.e., the predicates **P**) between classes are not precise, their quantification for a pair of class instantiations uses a continuous range $\langle 0, 1 \rangle$ based on fuzzy logic [13, 14], rather than with the traditional range $\{0, 1\}$. Every predicate between a pair of segments, using corresponding feature values from both the segments, computes a fuzzy truth value. An example of such a fuzzy predicate is *Is-Near*, which maps the coordinates of the centroids of two line segments to a fuzzy truth value. Formally, *Is-Near* $(i, j) = \exp(-\beta \times X_{i,j})$, where β is a constant scaling factor and $X_{i,j}$ is the distance between the segments i and j, computed geometrically. In addition to fuzzy predicates such as this, sometimes crisp predicates which map the feature values to $\{0, 1\}$ can also be useful. The predicates used in this study are shown in Table 8.1. Further, applicability of any predicate to a given pair of source and target class labels is determined by the semantic net. Table 8.2 shows the applicability of each of the predicates given in Table 8.1 in this case.

These predicates measure and reward the conformance of a given instantiation with respect to the heuristics implied by the knowledge base. We believe this to be a more natural approach, in contrast to penalty functions, especially in the context of weak constraints.

8.2.2 Representation

Typically, application of GAs to a problem requires some encoding scheme for transforming feasible solutions into representations amenable to GA search, binary string being the most common representation in practice. A "solution" for the labeling task is given by a list of labels, one for each segment in the image. Therefore, in this case, a simple vector of labels constitutes an adequate encoding scheme. For example, given seven segments, the vectors $\langle O, W, W, C, N, N, S \rangle$ and $\langle W, W, S, N, O, O, N \rangle$ are both feasible solutions. For the image in Figure 8.2, any vector $\langle C_1, C_2, \ldots, C_{35} \rangle$ such that $C_i \in \{N, S, W, C, O\}$ for all i, constitutes a feasible solution and, in turn, is a valid organism for GAs. By using

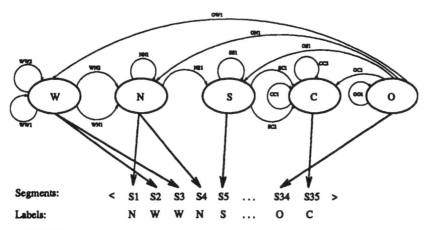

FIGURE 8.3
A simple semantic net for oceanic features.

sufficient number of bits at each allele position to encode any of the possible labels, such a vector of labels is encoded as a binary string suitable for GA search.

8.2.3   Fitness Function

The *Building Block Hypothesis* suggests that the power of GAs is derived from their ability to combine smaller building blocks of solutions together into better solutions. With this in mind, a fitness function in which each labeled segment can contribute directly to the fitness of the individual, in accordance with the propriety of that labeling, is highly desirable. The following bottom-up formulation of the fitness function addresses this issue. (An earlier top-down approach is described in Reference [1].)

Let n be the number of segments in the image. Consider any two segments S_r and S_t. Suppose they acquire the class labels C_i and C_j for the GA-generated individual whose fitness is being computed. This pair of segments, by inheritance from the semantic net, denotes a predetermined set of predicates (as shown in Table 8.2) which relates the class C_i to the class C_j. Let h_{ij} be the number of predicates from C_i to C_j. The fitness of the individual representing a complete labeling is computed as follows:

$$E = \sum_{i=1}^{n} \sum_{j=1}^{n} \sum_{k=0}^{h_{ij}} W_{ijk} \times P_{C_i C_j k}$$

$P_{C_i C_j k}$ is the kth predicate between class C_i and class C_j, computed as discussed earlier, W_{ijk} is the weight for this predicate, and h_{ij} is the total number of predicates applicable from C_i to C_j. In these experiments a uniform weight of 1.0 is assumed

Table 8.1 Description of Predicates

Name of the predicate	Formula	Comments
Is-North-Of(i,j)	If $AvgLat(i) \geq AvgLat(j)$, $= 1$; otherwise, $= 0$.	Segment i is north of segment j
Is-Near(i,j)	$\exp(-\beta \times X_{i,j})$	$X_{i,j}$ is distance between segments i and j
Is-Not-Near(i,j)	$1 - Is\text{-}Near(i, j)$	Fuzzy complement of *Is-Near*
*Is-North-Of-*And-*Fifty-Km-From(i,j)*	$\min \{ Is\text{-}North\text{-}Of(i, j),$ $\exp(-\beta \times \lvert X_{i,j} - 50\rvert) \}$	Example of a compound predicate built from simple predicates
*Arcs-Of-Circle-*And-*Less-Than-Hundred-Km-Distant(i,j)*	$\min \{ Arcs\text{-}Of\text{-}Circle(i, j),$ *Less-Than-Hundred-Km-Distant*$(i, j) \}$	*Arcs-Of-Circle* is estimated based on intersection of cords from segments; second predicate is computed as $= 1$, if $X_{i,j} \leq 100$, and $= \exp(-\beta \times \lvert X_{i,j} - 100\rvert)$, otherwise

for all predicates. It may be noted that all the predicates can be precomputed, since the number of segments and predicates is a constant for any given image. Since the actual computation of fitness for any individual during the evaluation step of the GA involves only the summation, this fitness function is computationally efficient.

8.3   Experimental Results and Analysis

Experiments were carried out using the representation and the fitness function described above. Splicer [2], a genetic algorithm environment, was used with modifications for these experiments. The parameters used in GA runs are shown in Table 8.3. Each run with these settings is repeated ten times, starting with a different initial random population each time. The results with respect to the fairly difficult infrared image shown in Figures 8.1 and 8.2 are described here. The accuracy of the best solutions generated by the GA in each run is compared with that of manual labeling. The results are given in Table 8.4. Figure 8.4 shows the best labeling obtained over these ten runs and Figure 8.5 illustrates the convergence

Table 8.2   Pairs of Class Labels with Applicable Predicates

Pred Id	Source i	Target j	Predicate name
P_{WW1}	W	W	*Is-Near(i,j)*
P_{WW2}	W	W	*Arcs-Of-Circle-And-Less-Than-Hundred-Km-Distant(i,j)*
P_{WN1}	W	N	*Is-North-Of(i,j)*
P_{WN2}	W	N	*Is-Near(i,j)*
P_{NN1}	N	N	*Is-Near(i,j)*
P_{NS1}	N	S	*Is-North-Of-And-Fifty-Km-From(i,j)*
P_{SS1}	S	S	*Is-Near(i,j)*
P_{SC1}	S	C	*Is-North-Of(i,j)*
P_{SC2}	S	C	*Is-Near(i,j)*
P_{CC1}	C	C	*Is-Near(i,j)*
P_{CC2}	C	C	*Arcs-Of-Circle-And-Less-Than-Hundred-Km-Distant(i,j)*
P_{OW1}	O	W	*Is-Not-Near(i,j)*
P_{ON1}	O	N	*Is-Not-Near(i,j)*
P_{OS1}	O	S	*Is-Not-Near(i,j)*
P_{OC1}	O	C	*Is-Not-Near(i,j)*
P_{OO1}	O	O	*Is-Near(i,j)*

Note: W = warm eddy; C = cold eddy; O = other; N = north wall of Gulf Stream; S = south wall of Gulf Stream.

of fitness values for that run.

These results are surprisingly good, considering the naiveté of the predicates used in these experiments. For example, analysis of the labeling found by the GA reveals that the majority of the mislabeled segments do not conform to the heuristics of the primitive domain knowledge encoded into the semantic net. In other words, the knowledge base was not detailed enough to guide the GA search correctly with respect to these segments. Second, unlike other successful approaches to this problem [8], results in this case were obtained from "cold-start" (*i.e.,* without using any information about the known last location of features) and, therefore, are all the more significant. Finally, this approach to labeling is not restricted to this single domain, but is applicable to a range of problems, where weak domain constraints can be appropriately quantified.

We believe that the performance of this approach can be strengthened a great deal in many ways. Obviously, additional predicates and more domain knowledge in computing the predicates are essential for describing the labeling constraints more accurately. Second, in the above, all the predicates are given equal weights in computing the fitness. However, it is desirable that the contribution of each

Table 8.3 Parameters for GA Runs

Description	Value
Population size	200
Number of generations	200
Selection operator	Proportional selection using sampling method of stochastic remainder with replacement
Crossover operator	Uniform crossover (allele level)
Probability of crossover	0.600
Mutation operator	Bit mutation
Probability of mutation	0.005

FIGURE 8.4
Best labeling by the GA.

predicate be in proportion to the strength of the heuristic represented by it. Third, in these experiments, all the segments in an image contribute equally to the fitness. However, certain properties, such as length, make some segments more important than others in an image, perhaps a better method would be to normalize the predicate values for each segment pair based on some of these properties.

Table 8.4   Accuracy of GA-Generated Labeling

Run #	Accuracy %	Run #	Accuracy %
1	80	6	63
2	57	7	71
3	66	8	77
4	83	9	71
5	83	10	69

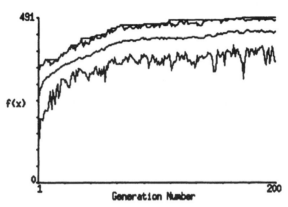

FIGURE 8.5
The convergence of best-ever, current-best, average, and worst fitness values.

8.4   Conclusions

In this work, feasibility of harnessing the power of GAs for the labeling task in remote sensing is demonstrated. The combination of using semantic nets to represent domain constraints and using fuzzy predicates to estimate the conformance of the labeling to these constraints has yielded a novel strategy for computing the fitness of solutions.

This methodology is currently being extended to a slightly different problem, namely, identification of cloud species from multispectral satellite images. A massively parallel implementation of this strategy is planned for effectively supporting large population sizes.

Acknowledgment

This work was supported in part by a grant from Naval Oceanographic and Atmospheric Research Laboratory, Stennis Space Center, MS, Grant # N00014-89-J-6003.

References

[1] Ankenbrandt, C.A., Buckles, B.P., and Petry, F.E., Scene recognition using genetic algorithms with semantic nets, *Pattern Recognition Lett.*, 11, 285, 1990.

[2] Bayer, S.E. and Wang, L., A genetic algorithm programming environment: splicer, in *Proc. 3rd Int. Conf. on Tools for Artificial Intelligence*, IEEE Computer Society Press, Los Alamitos, CA, 1991, 138.

[3] Buckles, B.P. and Petry, F.E., *Genetic Algorithms*, IEEE Computer Society Press, Los Alamitos, CA, 1992.

[4] Forrest, S., Jovornik, B., Smith, R.N., and Perelson, A.S., Using genetic algorithms to explore pattern recognition in the immune system, *Evol. Comput.*, 1(3), 191, 1993.

[5] Hill, A. and Taylor, C.J., Model-based image interpretation using genetic algorithms, *Image Vision Comput.*, 10(5), 295, 1992.

[6] Holyer, R. and Peckinpaugh, S., Edge detection applied to satellite imagery of the oceans, *IEEE Trans. Geosci. Remote Sensing*, 27(1), 46, 1989.

[7] Lea, S.M. and Lybanon, M., Finding mesoscale ocean structures with mathematical morphology, *Remote Sensing Environ.* 44(1), 25, 1993.

[8] Krishnakumar, N., Iyengar, S., Holyer, R., and Lybanon, M., Feature labelling in infrared oceanographic images, *Image Vision Comput.* 8(2), 142, 1990.

[9] Pal, S.K., Bhandari, D., and Kundu, M.K., Genetic algorithms for optimal image enhancement, *Pattern Recognition Lett.*, 15, 261, 1994.

[10] Seetharaman, G., Narasimhan, A., and Stor, L., Image Segmentation with Genetic Algorithms: A Formulation and Implementation, *Proc. SPIE Conf. on Stochastic and Neural Methods in Signal Processing and Computer Vision,* SPIE vol. 1569, San Diego, July 1991.

[11] Siedlecki, W. and Sklansky, J., A note on genetic algorithms for large-scale feature selection, *Pattern Recognition Lett.,* 10(11), 335, 1989.

[12] Van Hove, H. and Verscharen, A., Genetic Algorithms for Images, Technical Report, University of Antwerp, Belgium, 1994.

[13] Zadeh, L.A., Fuzzy logic, *Computer*, 21(4), 83, 1988.

[14] Zimmermann, H.J., *Fuzzy Set Theory and Its Applications*, Kluwer Nijhoff, Dordrecht, Netherlands, 1985.

9

Learning to Learn with Evolutionary Growth Perceptrons

Steve G. Romaniuk

Abstract   The ability to automatically construct neural networks is of importance, since it supports reduction in development time and can lead to simpler designs than traditionally handcrafted networks. Automation is further required to take the step toward a more autonomous learning system. In the first part of this chapter we outline an automatic network construction algorithm, which utilizes simple evolutionary processes to locally train network features using the perceptron rule. Emphasis is placed on determining the effectiveness of several types of crossover operators in conjunction with varying the population size and the number of epochs individual perceptrons are trained. In the second part, weaknesses of traditional *one-shot* learning systems are pointed out and the move toward an approach capable of displaying the ability to *learn about learning* is made. Trans-dimensional learning is introduced as a means to automatically adjust the *learning bias* inherent to all learning systems. Evidence will be provided showing for example, that learning the whole can be simpler than learning a part of it.

9.1 Introduction

Pattern recognition plays an influential role in many different fields of research, its application ranging from cognitive psychology to computational intelligence. Artificial intelligence and machine learning, neural networks, and evolutionary computation have over the past decades assigned significant importance to the area of pattern recognition.

The purpose of this chapter is to first outline a simple pattern recognition algorithm which is based on neural networks. The algorithm is self-configuring and utilizes the simple perceptron learning rule. This represents a pleasant alternative to various other approaches which are founded on complicated training algorithms and transfer functions. Evolutionary principles guide the selection of appropriate training sets during network construction. The second part of this chapter investigates several crossover operators in terms of their efficiency when evolving the necessary training sets. Finally, the third part investigates the applicability of this pattern recognition algorithm to the task of *learning to learn*. To this date very little research has been undertaken in the construction of algorithms that automatically adapt their inherent *learning bias* and display the ability to learn to learn from past experiences.

In order to utilize standard neural network learning algorithms such as back-propagation, several choices must be made by the user. These include, but are not limited to, number of hidden layers, number of hidden units, type of learning rule employed, choice of transfer function, and an assortment of learning parameters such as momentum term and learning rate. In general, determining an effective architecture for such multilayered feedforward networks can be both time-consuming and frustrating. Striving for optimality among potential network architectures in terms of hidden units and connections is, in general, of high priority, since deriving minimal configurations can curb total training time as well as the time required to test new patterns (during classification).

To this date several methods have been proposed to automatically construct neural networks and drive for reduction in network complexity, that is, determine the appropriate number of hidden units, layers, etc. [1, 2, 7, 9, 13, 14, 20, 26, 29, 31]. Many of these approaches utilize rather complicated algorithms and network structures. The simplest of these approaches is the Extentron [2] algorithm which is based on the simple perceptron rule [21]. Unfortunately, the algorithm is ineffective in determining suitable network structures for higher dimensional problems. For example, learning 2- and 3-bit parity requires a single hidden unit. Learning 5-bit parity already requires 5 hidden units, whereas for 6-bit parity the number shoots up to 11 hidden units. Other algorithms such as the Upstart [9] or Cascade Correlation [7] require about N hidden units to learn N-bit parity, but these algorithms are also increasingly more difficult.

By closely observing the perceptron learning rule for more difficult parity problems, it was noticed that by presenting the complete training set far fewer examples could be correctly recognized as opposed to removing a few prior to training. In other words, some of the examples interfered with the remaining ones, causing a sort of *confusion* within the perceptron to maximize recognition of all training patterns. This observation gave rise to further investigate the importance of providing the perceptron—at every stage of network construction—with the *right* set of training examples. The purpose for harnessing the perceptron rule lies in its simplicity. If we envision future hardware to execute existing training methods, then it is important to find algorithms as simple as possible that can do the job. The perceptron rule is probably the simplest of all weight adaptation algorithms.

In the following sections we give an outline of the Evolutionary Growth Perceptron (EGP) algorithm [29] and present some empirical findings regarding the choice of crossover operator and its significance, before moving on to outline how EGP can be used in an environment that requires the ability to learn about learning.

9.2 Constructing Networks with the Help of Evolutionary Processes

Genetic Algorithms (GAs) are in general simple adaptive search algorithms, which create new solutions to a given problem by exploiting past performance of older solutions in a manner similar to evolutionary processes as found in nature. The solution space from which the individual solutions are drawn is represented in form of a finite length string of chromosomes. Through use of genetic operators such as crossover and mutation, modifications to the individual chromosomes of a population are introduced in a systematic fashion based on optimizing some criteria. Binary encoding schemes of chromosomes have been popular in past GA use and are based on a fundamental theorem of GAs introduced by Holland [15]. According to Holland's schema theory binary coding schemes provide for low cardinality in obtaining a maximum number of similarities in what is encoded. Due to their simplicity and their past success in finding suboptimal solutions to NP-hard problems (such as Traveling Salesman [11, 12]), GAs seem ideal candidates for determining the *right* partitions of the training set during the various stages of network construction. Next, we point out how GAs can be utilized to develop effective partitions of the original train set.

Representing individual chromosomes of a population is straightforward. Assume we are given a set of training examples T. The size of each chromosome is then given by $|T|$ and presence/absence of a training example in the current partition is encoded by a binary 1/0, respectively.

The function to optimize is simply the number of examples of T that are correctly recognized by the current perceptron. We can express this more formally as

$$\mathcal{F}_{opt} = \max_{\forall C_i \in P} \left[O_{P_i}(T) \right] \tag{9.1}$$

where C_i is a chromosome (one particular partition of the original train set) and an element of the total population P. O_{P_i} is the ith perceptron trained on the partition of training examples represented by C_i.

The basic genetic operation made use of by the EGP algorithm can be summarized as follows:

- **Reproduction:** Half of the population is selected for further reproduction. The half includes those chromosomes that have the highest fitness according to the function which is optimized (as stated above).

- **Crossover:** This function creates new offspring from the original parents. In this particular implementation the parent chromosomes are grouped into pairs and from each of these pairs two new child chromosomes are created by a process of combination.

- **Mutation:** A mutation factor determines the rate at which a chromosomes' gene value is altered (toggled). Mutations add new information in a random way to the genetic search process, and ultimately help avoid GAs from getting stuck at local optimums. In the GA utilized in this chapter two parameters determine the rate at which chromosome's genomes are modified, that is, they determine how many chromosomes and how many genomes are changed during every generation.

9.3   Evolutionary Growth Perceptrons

The network constructed by EGP is similar to the ones generated by Cascade Correlation [7] and Extentron [2] algorithm. Only one hidden unit is created for each layer. Once a unit has been created it receives inputs from all previously generated hidden units, inputs and outputs. The later is in accordance with Divide & Conquer Networks [26] and represents a departure from the other two approaches. Since outputs are treated just like any other hidden unit, this requires training them one at a time. Once a hidden unit has been created, it is trained for a fixed number of epochs on a given partition of the train set. The unit's performance at the end of training is used to measure the chromosome's fitness. For each chromosome within the population a perceptron is trained. After the fitness has been determined the evolutionary operators of crossover and mutation are executed. The above outlined process is continued until all outputs have been trained. Figure 9.1 display high-level pseudo-code for the EGP algorithm.

$TRAIN\_OUTPUT()$:
(1) While $TRUE$ Do
    (1.1) Create Feature Cell C_F
    (1.2) $EVOLUTION(C_F)$
    (1.3) If $A(C_F, T) = |T|$ Then
        (1.4.1) Exit
(2) Halt

$EVOLUTION(\Rightarrow C_F)$
(1) $DETERMINE\_FITNESS(C_F)$
(2) Sort Population P by decreasing fitness
(3) $RecessionCnt = 0$
(4) For $MAX\_GENERATIONS$ Do
    (4.1) Perform Cross-Over on Population P
    (4.2) If $TimeToMutate$ Then
        (4.2.1) Perform Mutation of Population P
    (4.3) $DETERMINE\_FITNESS(C_F)$
    (4.4) Sort Population P by decreasing fitness
    (4.5) If $A(C_F, T) = |T| \vee RecessionCnt = MAX\_RECESSION$ Then
        (4.5.1) Exit
(5) Return

$DETERMINE\_FITNESS(\Rightarrow C_F)$:
(1) For all Chromosomes CH_i Do
    (1.1) For GEN_CYCLE_TIME Do
        (1.1.1) Train Feature Cell C_F using Perceptron Rule for one epoch
        (1.1.2) Measure Performance of C_F, Find $A(C_F, CH_i)$
        (1.1.3) If $A(C_F, CH_i) = \max_{\forall k, k<i}(A(C_F, CH_k))$ Then
            (1.1.3.1) Save Feature C_F for later use
            (1.1.3.2) If $A(C_F, CH_i) = |CH_i|$ Then
                (1.1.3.2.1) Exit
    (1.2) Restore saved Feature Cell C_F
    (1.3) Measure Performance of C_F, Find $A(C_F, T)$
    (1.4) $A(C_F, CH_i) = A(C_F, T)$
(2) Return

FIGURE 9.1
High-level description of EGP algorithm.

Nomenclature for EGP:

T:	set of training examples.
P:	set of population.
C_F:	feature cell.
CH_i:	ith chromosome of population P.
$A(C_F,T)$:	number of examples in T correctly classified by C_F.
MAX_GENERATIONS:	maximum number of generations allowed.
MAX_RECESSION:	number of generations for which there is no improvement, end evolution.
GEN_CYCLE_TIME:	maximum number of epochs perceptron is trained.

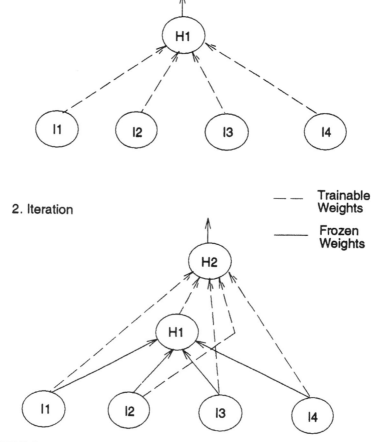

FIGURE 9.2
Parity-4 function as learned by EGP.

Figure 9.2 displays a representative network constructed by EGP for the 4-input even-parity function. During the first iteration, hidden unit H1 is recruited and it is trained using the perceptron rule. Once training is completed the unit's performance is measured. Since not all patterns are correctly recognized by unit H1, a new unit is installed (Iteration 2). Again training resumes for a fixed number of epochs. After including unit H2 all patterns characterizing the 4-input even-parity function are correctly learned and unit H2 is designated as an output.

9.4 Crossover Operators

9.4.1 Overview

The purpose of this section is to outline three approaches to constructing crossover operators for use by EGP.

1. **Simple crossover:** This widely applied operator utilizes a single crossover point which is selected by a random process.

2. **Weighted crossover:** Here the crossover point for each of the parents is chosen as a function of their relative fitness. The more fit one of the parent chromosomes is, the more of its original genetic code is passed on to its children. Crossover will always generate two new offspring. The ordering of genetic information is preserved during the crossover process. Crossover points between two parent chromosomes C_i and C_{i+1} are determined as follows:

$$CP_{C_i} = \frac{|T| * [|T| - F(C_i)]}{2 * |T| - [F(C_i) + F(C_{i+1})]} \qquad (9.2)$$

Here, CP_{C_i} represents the crossover point for chromosome C_i (calculated analogously for C_{i+1}). $F(C_i)$ indicates the fitness of chromosome C_i.

3. **Blocked crossover:** This approach attempts to create barriers between chunks of genes within a chromosome. These barriers act to prevent accidentally breaking up potential co-occurrences of genes that may have formed during the evolutionary process. By assigning a blocking factor to every location of a gene within a chromosome, a simple decision matrix (outline in Table 9.1) is used to decide whether a randomly selected crossover point for each of the parent chromosomes is a suitable choice. In Table 9.1 the column labeled P_i represents the randomly chosen crossover point for chromosome C_i. The $<$ and $\geq$ entries indicate whether the blocking factor associated with crossover point P_i is below a **blocking**

Table 9.1 Decision Matrix for Crossover Point P_i's Selection

P_{i-1}	P_i	P_{i+1}	Split
$<$	$<$	$<$	T
$<$	$<$	$\geq$	T
$<$	$\geq$	$<$	T
$<$	$\geq$	$\geq$	F
$\geq$	$<$	$<$	T
$\geq$	$<$	$\geq$	T
$\geq$	$\geq$	$<$	F
$\geq$	$\geq$	$\geq$	F

boundary or not. The final column contains a decision of whether a split point P_i can serve as a crossover point for chromosome C_i or not. For example, if the left and right neighbors of P_i (P_{i-1} and P_{i+1}) and P_i itself have blocking factors which are equal to or above the blocking boundary, a split does not occur (last entry of Table 9.1). In this case a new random split point is selected and the test repeated. If the test succeeds, P_i is chosen as a crossover point for chromosome C_i. Blocking factors for individual genes of a chromosome are maintained as simple counters. Whenever two parents mate, the blocking factor of a gene is incremented, if its current value (with respect to a gene's position in the chromosome) is preserved in the offspring after crossover has occurred; otherwise it is reset to 0.

For the following experiments we consider two values for the blocking boundary of the blocking method: blocking bounds 4 (B-4) and 10 (B-10).

9.4.2 EGP Experiments

In this section we evaluate EGP's performance to learn the well-known N-parity function, based on three different approaches for constructing crossover operators. The parity functions considered are 2- through 7-parity. In the experiments detailed in this section, ten trial runs were executed for learning each of the six parity functions. Figures 9.3 through 9.6 display the performance behavior for the different crossover operators. Results are provided that correlate the number of generation vs. the number of hidden units generated. Here, the label **S** refers to the simple operator, **W** to the weighted operator, and **B-n** to the blocked operator, with **n** being the blocking factor. The first four graphs are for the case MR = 20, where MR (max recession) indicates after how many unsuccessful generations (no improvement in error reduction) the evolutionary process is aborted.

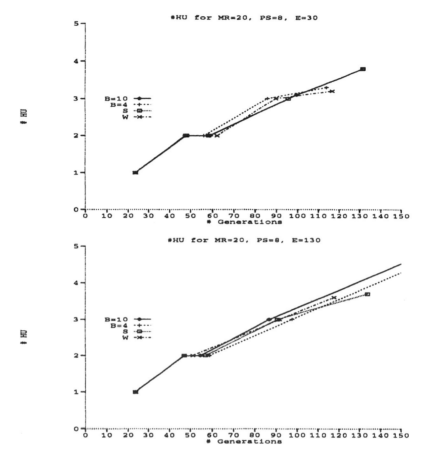

FIGURE 9.3
**# Generation vs. # hidden units for learning parity. MR = 20, PS = 8; and (a)
E = 30, (b) E = 130.**

For a population size PS of 8 and a maximum number of epochs E of 30 (Figure 9.3a), the weighted and blocked method (with a low blocking factor of 4) provides the best results in terms of reducing the number of hidden units and the number of generations required to derive at a solution. As the number of epochs is increased to E = 130 (Figure 9.3b), the weighted method again displays the best results, but the total number of generations required for the cases E = 30 and E = 130 is about the same, even though the training time for every individual perceptron is increased by more than a factor of 4. Incidentally, the same result holds for the other crossover operators. In fact, the B-4 operator yields a higher generation count for E = 130 than for E = 30. Also, the number of hidden units remains about the same, or is even slightly higher for E = 130. This indicates that for small population size, increasing the number of epochs may not translate into a

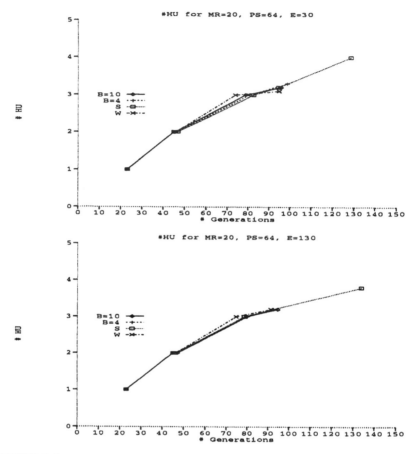

FIGURE 9.4
**# Generations vs. # hidden units for learning parity. MR=10, PS=64; and
(a) E=30, (b) E=130.**

decrease in the final network structure or the time required to obtain that structure.

As we increase the population size to PS = 64 (Figure 9.4) previous findings are completely reversed. First, for all crossover operators, except the simple operator, the total number of generations is significantly reduced from about 120 genera-tions (E = 30) to less than 100 generations. Again, similar to the previous case, increasing the parameter E has neither an effect on the final network structure nor the number of generations. Additionally, the difference in results between blocked and weighted crossover has diminished even further. This emphasizes that popu-lation size, if increased sufficiently, can offset differences in crossover operators (with the exception of the simple operator). On the other hand, for small popu-lation sizes the choice of crossover operator plays a more dominant role than, for example, the number of epochs used to train a perceptron. Finally, increasing PS

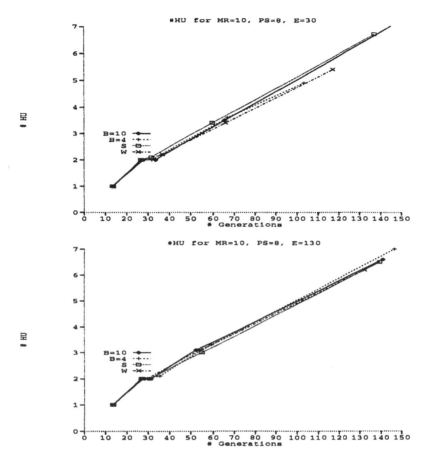

FIGURE 9.5
 # Generations vs. # hidden units for learning parity. MR = 10, PS = 8; and (a) E = 30, (b) E = 130.

also yields simpler network structures. For 6- and 7-parity the average number of hidden units drops from about 3.5 to 3.

For the second set of experiments we let MR = 10. We now observe for PS = 8 (Figure 9.5) a substantial change in the performance of various crossover operators. This time the difference between blocked B-4 and weighted is even more evident. Even though the number of generations is about the same as for the case MR = 20, we notice an apparent increase in the number of hidden units, regardless of the operator used. The hidden unit count jumps to about seven for 7-parity (B-10 and simple operator). Now, as the number of epochs is increased, we detect that the performance of the various crossover operators is almost identical. As a matter of fact, the B-4 and weighted operators actually experience a deterioration in their performance. From the combined results of the previous experiments we can

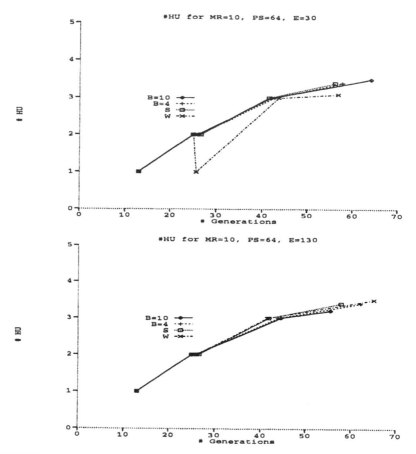

FIGURE 9.6
 # Generations vs. # hidden units for learning parity. MR = 10, PS = 64; and (a) E = 30, (b) E = 130.

conclude that increasing the number of epochs has little or even detrimental effects on EGP's performance. Here performance entails both the number of generations and number of hidden units required to learn N-parity.

By increasing the population size to 64 (Figure 9.6) we again note a substantial decrease in the number of generations (over 50%) for all operators compared to the case PS = 8. Also, similar to MR = 20, there is no apparent difference in performance of crossover operators. In all cases the final network architecture for 7-parity is around three hidden units. No apparent change in results is observed as E is increased.

We summarize our findings as follows: clearly, EGP strongly benefits from the size of its population. The choice of crossover operator only plays a role, if population size is relatively small. On the other hand, increasing the number of epochs for training a perceptron has no positive effects. We will utilize these findings in the next sections, when we address how EGP can be used in an environment which requires the ability to learn to learn.

9.5 Learning to Learn

9.5.1 Problems with Traditional Learning Systems

Over the last few decades a multitude of ideas have entered the realm of machine learning, resulting in many interesting approaches and advances to this important area. The oldest learning systems date back to the 1940s and the introduction of the perceptron model [21]. Later years would see the advances of inductive learning systems such as ID3 [25] (based on decision trees) or rule learning systems like AQ11 [23]. The 1980s saw a revival of the neural network community with the introduction of the backpropagation learning rule [30]. More recently approaches have been formalized that attempt to integrate seemingly different paradigms such as connectionism and symbolism [17, 24, 27, 33, 34]. Other advances have been made to automate the network construction process [2, 7, 9, 26].

With the discovery of ever new methodologies to tackle learning problems, the question of evaluating these diverse algorithms has remained mostly unchanged. In general we can identify two approaches: the first approach considers an empirical study which includes testing a new learning algorithm $L \in \mathcal{L}$ (element of the class of all learning algorithms $\mathcal{L}$) on a set of either artificial or natural (real world) problems $P \in \mathcal{P}$. This set P, of course, is finite and, in fact, constitutes as little as one to about a dozen problems. In general, several accepted tasks—also known as benchmarks—are selected for an empirical study. Besides including a set of benchmark problems, a few well-known learning algorithms (well publicized) may be selected in order to perform a comparison. Evaluation of learning algorithms proceeds by dividing a set of patterns that describe the problem under consideration into two partitions: a train (T_{Train}) and test set (T_{Test}). The degree of generalization (performance accuracy) of algorithm L is measured by determining how many patterns in T_{Test} are correctly recognized by L after being trained on patterns in T_{Train}. After training and testing have been completed for a given problem P, results obtained thus far are removed and a new problem is tackled. We refer to this procedure as *one-shot* learning.

It is a well-known fact that such an empirical comparison cannot yield any significant results due to the sheer size of potential problems that can exist in the real world. In fact, it is quite easy to verify that for every problem $P = T_{Train} \cup T_{Test}$, for which some algorithm $L \in \mathcal{L}$ can achieve accuracy α (denoted by $A_L[T_{Test}] = \alpha$), there always exists a problem $P' = T_{Train} \cup T'_{Test}$ such that $A_L(T'_{Test}) = 1 - \alpha$. This fact is true regardless of how we partition the examples describing problem P into train and test sets, as long as $T_{Train} \cap T_{Test} = \emptyset$.

The above essentially states that there cannot exist a universal learning algorithm (neither biological nor artificial) which can perform equally well on all classification problems. Instead, every learning algorithm must exploit some *learning bias* which favors it in some domains, but handicaps it in others. The following may serve as an example: Both decision tree-based as well as backpropagation-like learning algorithm have a more difficult time learning highly nonseparable functions (e.g., parity) than a truly linear function (e.g., logical or). When we say *more difficult to learn*, we refer to training time, as well as how the function is represented and the performance accuracy we can expect after training has occurred. A decision tree-based algorithm, for example, grows a more deeper tree structure, whereas a backprop network requires more hidden units (and connections) to learn even/odd parity. Due to these observations, it should become apparent why simple empirical studies are of little value in deciding the quality of a new learning approach or allowing for comparing different approaches.

A second approach to evaluating learning systems is based on developing a theoretical model. A good example of such an approach is the *probably approximately correct* learning model (PAC for short) [36]. In this approach, one attempts to identify subclasses of problems which can be learned in polynominal time and for which confidence factors can be stated for achieving some performance accuracy. Unfortunately, these methods appear to be too rigid and constrained to be of any use. For one, subclasses may be very specific, such that results about them seem to say very little. A more substantial shortcoming is subclasses of problems tend to be highly abstract. For example, a restricted class of problems might be $CNF_n(k)$, that is, all propositional formulas over n boolean variables in conjunctive normal form with at most k literals. Results like this do not outline how abstract results can be placed in relation with real world problems. Having determined that algorithm L tends to fare well with problems belonging to some subclass of problems still does not help us decide how to solve specific real world problems, unless we show that the model class is in some correspondence with the real world, but to determine this may require utilizing algorithm L in the first place.

The problem that we are facing—to develop learning algorithms that can be employed to solve various real world problems—cannot be overcome with simple *one-shot* learning algorithms. For this matter, we propose to look at algorithms which have the capability to *learn about learning*. These new algorithms must be capable of shifting their built-in bias (to favor certain subclasses of problems) automatically, by utilizing past performance, to give them a wider applicability than simple one-shot learning systems. In the next section we outline one such approach, which is based on trans-dimensional learning.

9.5.2 Trans-Dimensional Learning

The ability to learn knowledge on top of already existing knowledge without having to restart the learning process from ground zero can affect learning time, representational complexity, as well as generalization on yet unseen patterns. To move beyond simple *one-shot* learning we consider the following learning task:

Given: A set of problems $\mathcal{P} = \{P_1^{n_1}, P_2^{n_2}, \ldots, P_k^{n_k}\}$, where $P_i^{n_i} \in C^{n_i}$. The $P_i^{n_i}$ belong to the class of classificatory problems C (continuous inputs/binary output) of variable input dimension n_i.

Goal: Develop a network N_k that can correctly classify all $P_i^{n_i} \in \mathcal{P}$. No specific order is imposed on the problems in $\mathcal{P}$ when presented to the network constructing algorithm.

We refer to a network constructing algorithm that can solve the above problem as a *trans-dimensional learner*. We should note at this point that a *network constructing* trans-dimensional learner is one of many possibilities. For example, we could have chosen a decision tree-type representation, instead of a connectionist network for constructing TDL. A choice in favor of the connectionist representation was made, due to earlier obtained positive results involving EGP and considering the algorithm's simplicity.

In the following sections, we outline some of the basic requirements that are essential for our choice of TDL. Specifically, we address the difficulty of feature construction to solve individual problems in $\mathcal{P}$, local feature pruning to reduce network complexity, and finally feature selection to curb learning time and improve recognition.

9.6 Local Feature Pruning during Network Construction

9.6.1 Why Employ Feature Pruning?

Clearly, sharing cells and constructing networks in the manner Divide & Conquer Networks [26], Cascade Correlation [7], Extentron [2], and other network growing algorithms proceed with, can lead to a large number of connections for hidden and output cells (hence referred to as feature cells). Not only can the fan-in to every feature cell increase drastically (as the size of network increases), but the overall network complexity in terms of connections can also grow out of bound. Constraining this excessive growth is eminent both from a hardware and software point of operation, since it will directly affect learning time and classification time during testing. Second, it has been pointed out that generalization can improve as a result of network pruning [18, 27]. Finally, pruning determines the importance of connections in acting as good feature detectors and may help in the knowledge extraction process [27]. All three points combined should make it clear why *effective* pruning of networks is of such importance.

MULTI_PASS_PRUNING():
(1) OK = FALSE
(2) For all $C_{i,j}$ of Feature C_{F_i} do
    (2.1) $I(C_{i,j}) = FALSE$
    (2.2) $OriginalW_{i,j} = W_{i,j}$
    (2.3) $W_{i,j} = 0$
(3) While $\neg OK$ do
    (3.1) OK = TRUE
    (3.2) For all $I_k \in T$ do
        (3.2.1) Perform single feedforward pass.
        (3.2.2) If $\|O_i(I_k) - D(I_k)\| > \delta$ then
            (3.2.2.1) $UPDATE\_CONNECTION()$
            (3.2.2.2) $OK = FALSE$
    (3.3) If $\neg OK$ then
        (3.3.1) RETURN_BEST_LINK()

UPDATE_CONNECTION():
(1) For all $C_{i,j}$ of Feature C_{F_i} do
    (1.1) If $C_{i,j} = 0$ then
        (1.1.1) $\epsilon = \|O(A_i + OriginalW_{i,j} * O_j(I_k)) - D(I_k)\|$
        (1.1.2) If $\epsilon < \epsilon_{Min}$ then
            (1.1.2.1) $\epsilon_{Min} = \epsilon$
            (1.1.2.2) $Input_{Best} = j$
(2) Increment $IC(C_{i,Input_{Best}})$

RETURN_BEST_LINK():
(1) For all $C_{i,j}$ of Feature C_{F_i} do
    (1.1) If $W_{i,j} = 0 \wedge IC(C_{i,j}) > IC_{Max}$ then
        (1.1.1) $IC_{Max} = IC(C_{i,j})$
        (1.1.2) $j_{max} = j$
(2) $I(C_{i,j_{max}}) = TRUE$
(3) $W_{i,j_{max}} = OriginalW_{i,j_{max}}$

FIGURE 9.7
Multipass connection pruning algorithm.

9.6.1.1 Multipass Pruning

A high level algorithmic description is presented in Figure 9.7. At the heart of this algorithm is the desire to successively identify connections from the original feature F_i, which have the largest impact on recognizing training examples $I_k \in T_{Train}$ and remove those deemed redundant. To accomplish this feat, a copy of the original weight vector is maintained in $OriginalW_i$ for feature F_i, along with a binary importance measure $I(C_{i,k})$ and an importance count $IC(C_{i,k})$ of every connection $C_{i,k}$.

Initially, the importance measure is set to false for every connection and the weight vector of the current feature is saved and then zeroed out (step 2, *MULTI_PASS_PRUNING*). The purpose of the zeroing out is to emphasize that initially all connections are deemed unimportant. Now, as long as the new feature does not correctly recognize those training examples in T_{Train}, which the original feature F_i did, each connection weight stored in *OriginalW*$_i$ is considered a potential candidate for inclusion into the new feature. To decide which connection to select requires measuring the connections' importance in maximizing the recognition rate of training patterns $I_k \in T_{Train}$. The function *UPDATE_CONNECTION* determines from all connections not yet deemed important of the original feature F_i those that best reduce the error difference between the actual and desired output (ϵ) for the training pattern I_k currently under consideration. The connection which delivers the smallest error difference ϵ has its importance count incremented (step 2, *UPDATE_CONNECTION*). Finally, after one iteration of the loop in step 3 of *MULTI_PASS_PRUNING* the connection with the highest importance count is returned (step 3.3) by having its original weight copied back into W_i (function *RETURN_BEST_LINK*). The loop in step 3 is executed until all patterns recognized by the original feature (step 3.2.2) are also recognized by the new feature.

The process is guaranteed to come to a halt after all (initially) deemed unimportant connections have their status changed to important. Thus, the maximum number of iterations (or passes) required is bounded by K, the number of connections to the current feature F_i.

Multipass pruning can be considered a bottom-up approach in that connections are selected based on their importance in minimizing the error difference between current feature output and desired output. This stands in contrast to other pruning approaches which apply a top-down approach, where one or more connections are pruned from the set of all connections. In Reference [28] an empirical performance analysis involving multipass pruning (applied to Divide & Conquer Networks) and *Optimal Brain Damage* [18] is presented for several real world domains.

Nomenclature for multipass pruning:

C_{F_i}:	*ith feature cell.*
I_k:	*kth training example.*
$C_{i,k}$:	*kth connection of ith feature cell.*
$W_{i,k}$:	*associated kth connection weight for ith feature cell.*
OriginalW$_{i,k}$:	*original kth connection weight for ith feature.*
$O_i(I_k)$:	*output for ith feature cell for training example I_k.*
A_i:	*cell activation for ith feature cell.*
$I(C_{i,k})$:	*importance of kth connection for feature cell C_{F_i} (Boolean value).*
$D(I_k)$:	*desired output for training example I_k.*
δ:	*maximum absolute error difference allowed between feature C_{F_i} and the desired output for example I_k.*
$O(a)$:	*output transfer function for activation a.*
$IC(C_{i,k})$:	*importance count of kth connection of ith feature C_{F_i}.*

9.7   Feature Selection

Two major shortcomings of constructive algorithms such as EGP, Cascade Correlation, and Extentron are that besides allowing shortcut connections from input to hidden units, all hidden units are connected to every newly created hidden unit and output. This causes an increase in the number of total connections in a network as opposed to a more conventional architecture (such as Dynamic Node Creation [14], Hirose's algorithm [14]), which adds hidden units to a single layer (horizontal growth only). Second, only one hidden unit is added per layer. This can very quickly lead to an increase in layers, a rather undesirable effect, which can be detrimental to parallel execution, since an activation impulse will have to travel through all preceding layers before reaching the current layer. This stands in contrast to networks that allow multiple hidden units per layer and thereby support a higher degree of parallelism by reducing the number of hidden layers (e.g., Divide & Conquer Networks [26]).

To address these problems for TDL—especially, since EGP forms the basic network construction algorithm utilized—emphasis has to be placed on effectively selecting hidden features (units) when constructing new features. We address the merits of two such approaches:

1.   Single hidden feature selection

2.   Multiple hidden feature selection

The first approach is both obvious and simple. After having learned problems $P_1, P_2, \ldots, P_k$ the network constructed so far is described by network N_k. Now, assume an attempt by N_k to provide 100% correct recognition if problem P_{k+i} fails. As a consequence of this, a new feature is created. Initially, connections are provided from all input features to $C_{F_{k+1}}$. Then, every feature in N_k is evaluated according to its fitness to correctly recognize P_{k+i}. This recognition requires a single presentation of all training patterns in T_k to N_k. Selection then simply proceeds by picking the feature in N_k that has the highest recognition rate. If more than one such feature exists, the one which is positioned on the lowest layer is selected. Again, if more than one can be identified, an arbitrary selection takes place. Selecting the feature on the lowest layer is incorporated, in order to curb unnecessary vertical growth of network N_k. On the other hand, picking only one feature curbs connection growth of the network. Unfortunately, selecting a single feature from the previously constructed network can also be detrimental. It is quite conceivable that problem P_{k+i} could be solved more readily utilizing several low-level features by forming an appropriate linear combination of them.

To overcome some of the problems for single-feature selection, a quality measure is associated with every feature C_{F_i}. Selection now proceeds by identifying all those features which have a measure of quality exceeding some threshold α.

This measure is based both on quality of recognition and a cost penalty, which is determined by a feature's layer location with respect to the depth of network N_k. The quality measure is given by

$$
Q_{k,i}\left(C_{F_i}, P_k\right) = \frac{A_k\left(C_{F_i}, P_k\right)}{|T_k|}\left(1 - \frac{\log\left[L\left(C_{F_i}\right) + 1\right]}{\log\left(\max_{\forall j}\left[L\left(C_{F_j}\right)\right] + 1\right)}\right) \tag{9.3}
$$

where $L(C_{F_i})$ returns feature C_{F_i}'s layer location. The first factor of Equation 9.3

(1) For all P_k do

    (1.1) While $\frac{A_k(C_{F_{Current}}, P_k)}{|T_k|} < 1$ do

        (1.1.1) Create new feature $C_{F_{new}}$

        (1.1.2) Connect all input units to $C_{F_{new}}$

        (1.1.3) For all $C_{F_i} \in N_k$ do

            (1.1.3.1) If $Q_{k,i}(C_{F_i}, P_k) > \alpha$ then

                Connect C_{F_i} to $C_{F_{new}}$

        (1.1.4) $N_{k+1} = N_k \cup \{C_{F_{new}}\}$

        (1.1.5) $C_{F_{Current}} = C_{F_{New}}$

        (1.1.6) EGP(N_{k+1})

        (1.1.7) If FeaturePrune then

            (1.1.7.1) $MULTI\_PASS\_PRUNING(C_{F_{Current}})$

FIGURE 9.8
TDL algorithm with multifeature selection and feature pruning option.

measures feature C_{F_i}'s accuracy on problem P_k, whereas the cost factor incorporates information about feature C_{F_i}'s layer location relative to network N_k's height. Figure 9.8 depicts a high-level description of the multiple-feature selection procedure in conjunction with EGP and optional feature pruning. In order to deal with training patterns of various input dimensions, a pool of input features is initially set aside. The pool is left-adjusted, which means that the first input value of a training pattern is stored in the first input feature unit of the pool and so on, until all input values have been assigned to input feature units of the network. Remaining feature units are assigned a default value of 0, to guarantee that no activations are propagated from any potential connections that emanate from these units to other network feature units (same effect as being disconnected).

Finally, in Figure 9.9 an example of a TDL-generated network is presented. The network correctly recognizes the parity- and or-function of input dimensions 2 to 5 (note that pruning has been applied).

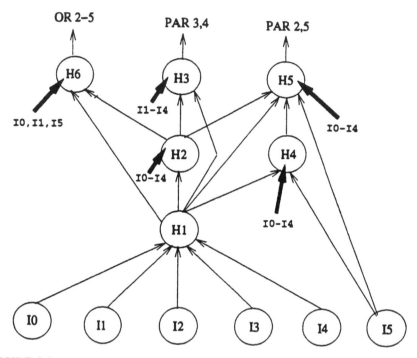

FIGURE 9.9
PARITY- and OR-functions 2 to 5 as learned by TDL.

9.8 TDL Experiments

9.8.1 Outline of Experiments

After having outlined the basic TDL configuration, we now present several empirical findings. In order to study TDL's ability to efficiently grow networks, we look at several performance parameters.

- **Number of generations:** To determine the time it takes to learn some problem class of trans-dimensional learning examples. This performance parameter forms the basis for comparison with all subsequently listed measures.

- **Number of combined units:** First measure to determine complexity of derived networks. We consider **number of combined units** to represent all units in the network which are not input units.

- **Number of connections:** Second measure used in determining the complexity of derived networks. Change over number of generations of training time is pointed out.

- **Number of layers:** A peripheral measure for determining network complexity. Of importance during forward-pass of yet unseen examples; solely for classification.

In order to obtain more representative performance measures, average results compiled over ten trial runs are reported for all experiments. Performance measures are placed in relation with the number of generations required to create individual networks. In all experiments the population size utilized by EGP is varied between 8 and 64, and the maximum number of cycles each perceptron is trained is fixed at 60 epochs. To obtain an upper bound on the number of perceptron training cycles (assuming serial execution) simply requires forming the product of number of generations with a factor of either 8 or 64 multiplied by 60.

9.8.2 Domains of Study

In order to form trans-dimensional problems the following frequently employed benchmarks have been included in this study:

- **And-functions:** Considered are all functions of input size 2 through 7.

- **Or-functions:** Again inputs range from 2 to 7.

- **Even- and odd-parity:** Requires output to be set based on number of active inputs to either form an even or odd number of 1s, a highly nonseparable problem. Input dimension is chosen between 2 and 7.

- **Encoder:** The task is to associate an input pattern with an identical output pattern. Encoder problems considered here are 4, 8, and 16.

- **Adder:** Perform binary addition of two N bit numbers represented as a single binary input pattern. Cases considered here are addition of 2- and 3-bit numbers.

Combining the above problems results in a pool of 29 functions which have been selected for investigation (29-fkt problem).

9.8.3   Experimental Results

9.8.3.1   Learning Even/Odd-Parity

Learning even/odd-parity is depicted in Figures 9.10 through 9.12. Throughout all experiments the number of epochs is fixed at E = 60. This assignment is due to earlier findings, which suggested that the number of epochs plays little significance in reducing the number of generations and the number of combined units required to represent all 12 parity functions. Here, combined units is defined as the number of all hidden and output units of a network.

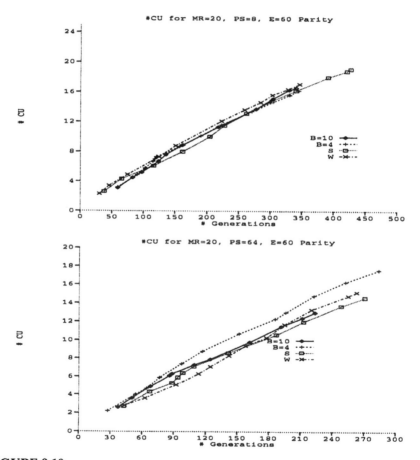

FIGURE 9.10
 # Generations vs. # combined units for learning parity. MR = 20, E = 60; and (a) PS = 8, (b) PS = 64.

For PS = 8 and MR = 20 (Figure 9.10) all crossover operators except the simple operator provide very similar results. Each of the marks on a graph indicate when a parity problem has been learned by TDL. From the distribution of marks for the

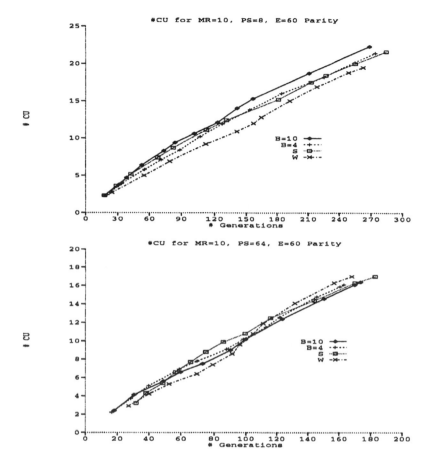

FIGURE 9.11
 # Generations vs. # combined units for learning parity. MR = 10, E = 60;
and (a) PS = 8, (b) PS = 64.

weighted operator (Figure 9.10a), we can observe that about half of the original 12 parity functions are learned in the final one third of training time (generations 260 to 350). This indicates that TDL—after being exposed to several instances of similar functions—has learned to more rapidly recognize other parity functions, resulting in a decrease in training time for learning the remaining functions.

As PS is increased to 64 (Figure 9.10b), the most striking observation we can make is the apparent reduction in training time. This result mirrors similar earlier findings, when EGP was utilized for learning individual parity problems (one-shot learning). A major difference to previous results is a noticeably better performance by the blocked method B-10 compared to other operators. Training time drops from about 350 generations to 230, whereas the number of combined units dips from around 15 to 12. This result is intriguing since the number of combined units

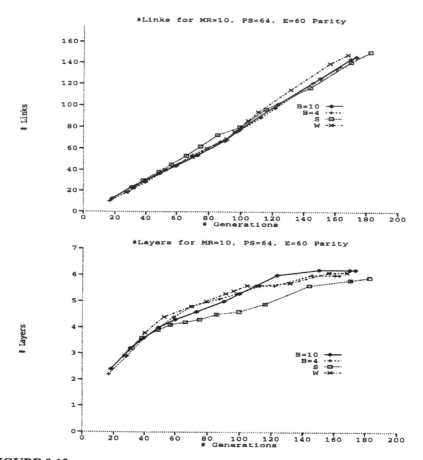

FIGURE 9.12
 # Generations vs. (a) # connections and (b) # layers for learning parity. MR = 10, E = 60, PS = 64.

(hidden + output) is identical to the number of parity problems learned (even + odd).

Even though five of the final parity problems are learned in the later stages of the training phase, the difference is not apparent as for the PS = 8 case. Nonetheless, it can be observed for the weighted and the blocked operator B-4.

For the next set of experiments we let MR = 10. The first observation we can make is the drop in generations from 350 (MR = 20) to about 270 (Figure 9.11a). Also, there is very little difference between the crossover operators when it comes to the total number of generations. On the other hand, we notice a clear increase in the number of combined units: average increase from 16 to 20.

As PS is increased to 64 (Figure 9.11b) the number of generations further dips from a previous low of 270 to about 170. Combined units also drop to around 16.

Unfortunately, this time no clear low of 12 units—as observed for the first case—is encountered for any of the crossover operators.

Finally, we report graphs indicating the number of connections (Figure 9.12a) and layers (Figure 9.12b) required for the case MR = 10 and PS = 64. The graphs displayed for the number of connections are almost identical in appearance to the ones earlier stated for combined units (linear growth with regard to the number of generations). On the other hand, the number of layers initially grows faster but rapidly flattens out about half way through the training session. In case of the weighted operator, the flattening seems to coincide with the point at which about half of all parity problems have been learned.

9.8.3.2 Learning 29-fkt Problem

In this final experiment we investigate TDL's performance on the 29-fkt problem. Note, that the earlier studied parity problem is a subset of the 29-fkt problem. For the case MR = 20 and PS = 8 (Figure 9.13a) we obtain most favorable results (least number of generations and combined units) for the weighted and the blocked B-10 operator. This is in correspondence with earlier obtained results involving the parity problem. Three facts stand out when compared to the previous experiment: first, there is a significant difference in the performance of the four crossover operators. Differences in training time are over 120 to 150 generations when compared to the simple and B-4 operator. Second, the response curves are not linear but display a sudden steep increase in the final stages of the training process, as opposed to the linear response obtained during learning the parity problem. Third—and most important—for the blocked B-10 operator around 20 functions (of the original 29) are learned in the final stages of training, that is, in the last 50 generations. For the weighted operator this result is obtained within 30 generations. In other words, more than two thirds of the problems are learned in the last one fifth of total training time. This result underlines how TDL benefits from previously learned information and indicates how the ability to learn to learn (from previously presented problems) has profound effects on the final results.

The most impressive finding can be gleaned, when comparing the absolute magnitude in number of generations required to learn either parity or 29-fkt problem. The parity problem is learned after about 350 generations have passed, whereas the 29-fkt problem is learned after only 280 generations (weighted and B-10 operator). This finding indicates that the time needed to learn a subset of functions can be reduced by learning additional functions, even when they have little in common with the original set of functions. Recall the composition of the 29-fkt problem: besides containing the simpler and- and or-functions, it also consists of encoder and adder functions. The task of identifying even and odd parity has little in common with the task of performing binary encoding or binary addition. This result supports the claim that in TDL learning the whole can sometimes be simpler than learning just a part of it.

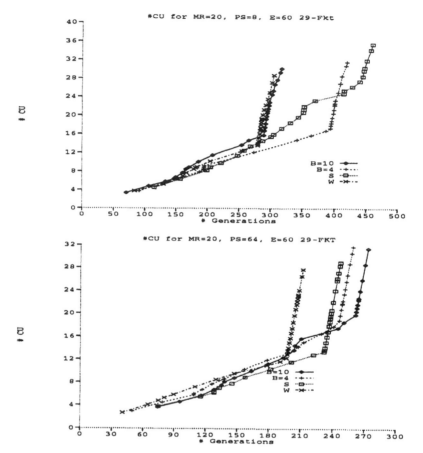

FIGURE 9.13
**# Generations vs. # combined units for learning 29-fkt. MR = 20, E = 60;
and (a) PS = 8, (b) PS = 64.**

For the next experiment we increase PS to 64 (Figure 9.13b). Compared to the
previous case of PS = 8, we observe that the overall training time substantially
decreases for all crossover operators (a result observed in previous experiments).
The simple operator appears to benefit the most from an increase in population
size. It experiences a drop from more than 450 generations to about 250. Also, the
number of combined units decreases from more than 32 to about 28 (for weighted
operator). Again, this is an intriguing result, since the number of combined units
is less than the number of functions learned. Both for parity as well as the 29-fkt
problem, the increase in number of combined units with respect to the number of
functions learned (regardless of the problem's input dimension) is **linear**. Finally,
the time frame in which about two thirds of the functions are learned has been
reduced to a mere 20 generations (less than 10% of total training time).

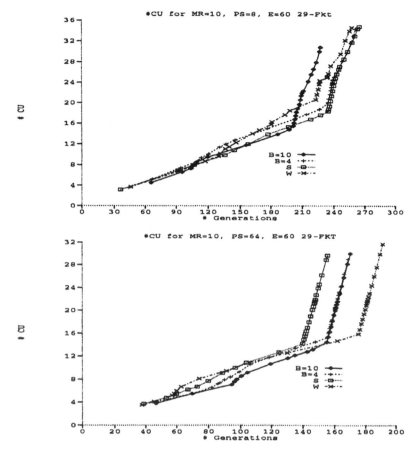

FIGURE 9.14
 # Generations vs. # combined units for learning 29-fkt. MR = 10, E = 60;
and (a) PS = 8, (b) PS = 64.

As MR is reduced to 10 (Figure 9.14) we note that the performance of all four crossover operators is more similar than for the MR = 20 case. Also, the overall training time is reduced, whereas other performance results remain unchanged. After increasing PS to 64 (Figure 9.14b) we observe another dip in the number of generations and the average number of combined units. Most impressive is the exceptional behavior of the simple crossover operator. With it TDL only requires about 150 generations to learn the 29-fkt problem.

The curves reflecting the increase in the number of connections (Figure 9.15a) are almost identical to the ones obtained for the combined units. Figure 9.15b depicts the number of generations vs. number of layers. Apparently, the final number of layers is about the same for the various crossover operators. The steep increase in layers coincides with the increase in combined units. This indicates that

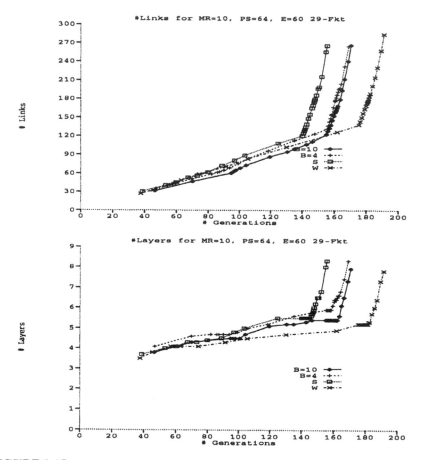

FIGURE 9.15
 # Generations vs. (a) # connections and (b) # layers for learning 29-fkt. MR = 10, E = 60, PS = 64.

newly formed feature detectors in the network are constructed from already existing low-level feature detectors. In other words, overall savings in combined units and training time are realized by creating high-level feature detectors, resulting in an overall increase in the number of layers.

Finally, in Figure 9.16a the accuracy (degree of generalization on yet unseen patterns) of TDL for the 29-fkt problem is displayed for the case MR = 10 and PS = 64. After presenting a single function, more than 70% accuracy is obtained. Note the graphs reflect the accuracy TDL obtains when tested on all 29 functions. Figure 9.16b reports performance for the simple crossover operator (denoted by actual S) and the expected performance. The expected curve is obtained by measuring during every presentation of a new function to TDL the total number of patterns it has been trained on. Hence, the difference between the two curves reflects the

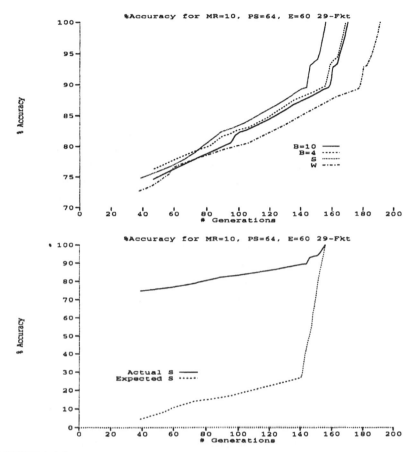

FIGURE 9.16
Generations vs. % accuracy for 29-fkt. MR = 10, E = 60, PS = 64; and (a)
actual, (b) actual and expected for S operator.

actual degree of generalization on yet unseen patterns. Reviewing Figure 9.16a,
we note that the response curves are very similar in nature. For the simple operator
more than 75% of all patterns are correctly identified after a single presentation of
a function belonging to the 29-fkt set. Similar to the earlier observed steep increase
in combined units, accuracy also picks up significantly. Clearly, TDL obtains a
high degree of generalization early on and maintains it through most of the training
process.

9.9 Summary

Taking the step from one-shot learning to develop an automatic network construction algorithm capable of trans-dimensional learning can be seen as a step toward a new and more powerful learning environment. Viewing units within a network solely as features and learning as a process of bottom-up feature construction were necessary notions to develop a feasible implementation of a trans-dimensional learner. Basing local feature training on the simple perceptron rule, combining evolutionary methods to effectively create training partitions, and pointing out the relative advantages of various crossover operators have substantially contributed toward the construction of a more flexible learning system and its understanding. Benefits obtained from employing TDL included: observations such as network complexity reduction in terms of combined units, connections, and layers. Contrary to intuition, it was noted that learning a large set of diverse problems can be either equal to or even simpler than attempting to learn a subset of the very same problems. Even across a highly diversified set of domains can powerful hidden features be constructed and help decrease learning time and network complexity as more problems are encountered. These powerful hidden features support adjusting *learning bias* depending on the type of problems presented by simplifying the learning process itself and help elevate TDL above *one-shot* learning systems. The findings of this study are of importance, since they suggest that learning new features can be substantially improved by learning on top of preexisting knowledge, even if there appears little in common between the two. Even though this is an initial study in the applicability of trans-dimensional learning to solve the general problem of *learning about learning*, it is one that has given rise to some interesting results and it is hoped that future work will prove as fertile.

References

[1] Ash, T., *Dynamic Node Creation in Backpropagation Networks*, Technical Report ICS Report 8901, Institute Cognitive Science, University of California, San Diego, La Jolla, CA, 1989.

[2] Baffes, P.T. and Zelle, J.M., Growing layers of perceptrons: introducing the extentron algorithm, in *Proc. 1992 Int. Joint Conf. on Neural Networks*, Baltimore, MD, June 1992, II-392.

[3] Beale, R. and Jackson, T., *Neural Computing: An Introduction*, IOP Publishing, New York, 1991.

[4] Chandrasekaran, B., Goel, A., and Allemang, D., Connectionism and information—processing abstractions, *AI Mag.*, Winter 1988.

[5] Fahlman, S.E. and Hinton, G.E., Connectionist architecture for artificial intelligence, *IEEE Comput.*, 1987.

[6] Fahlman, S.E., Faster-learning variations on back-propagation: an empirical study, in *Proc. 1988 Connectionist Models Summer School*, Morgan Kaufmann, San Mateo, CA, 1988.

[7] Fahlman, S.E. and Lebiere, C., The cascade-correlation learning architecture, in *Advances in Neural Information Processing Systems 2*, Touretzky, D., Ed., Morgan Kaufmann, San Mateo, CA, 1990, 524.

[8] Feldman, J.A., Dynamic connections in neural networks, *Biol. Cybern.*, 46, 1982.

[9] Frean, M., The upstart algorithm: a method for constructing and training feedforward neural networks, *Neural Comput.*, 2, 198, 1991.

[10] Freeman, J.A. and Skapura, D.M., *Neural Networks, Algorithms, Applications and Programming Techniques*, Addison-Wesley, Reading, MA, 1991.

[11] Grefenstette, J.J., Gopal, R., Rosamita, B., and Gucht, D.V., Genetic algorithms for the traveling salesman problem, in *Proc. Int. Conf. on Genetic Algorithms and Their Application*, Lawrence Erlbaum & Associates, NJ, 1985, 160.

[12] Grefenstette, J.J., Incorporating problem specific knowledge into genetic algorithms, in *Genetic Algorithms and Simulated Annealing*, Davis, L., Ed., Morgan Kaufmann, Los Altos, CA, 1987.

[13] Hall, L.O. and Romaniuk, S.G., A hybrid connectionist, symbolic learning system, *AAAI-90*, Boston, MA, 1990.

[14] Hirose, Y., Koichi, Y., and Hijiya, S., Back-propagation algorithm which varies the number of hidden units, *Neural Networks*, 4, 61, 1991.

[15] Holland, J.D., *Adaption in Natural and Artificial Systems*, University of Michigan Press, Ann Arbor, 1975.

[16] Honavar, V. and Uhr, L., A network of neuron-like units that learns to perceive by generation as well as reweighting of its links, in *Proc. Connectionist Summer School*, Morgan Kaufmann, San Mateo, CA, 1988.

[17] Magrez, P. and Rousseau, A., A symbolic interpretation for back-propagation networks, *Int. J. Intell. Syst.*, to appear.

[18] Le Cun, Y., Denker, J.S., and Solla, S.A., Optimal brain damage, in *Advances in Neural Information Processing Systems 2*, Touretzky, D., Ed., Morgan Kaufmann, San Mateo, CA, 1990, 598.

[19] Lim, S.F. and Ho, S.B., Dynamic Creation of Hidden Units in Backpropagation, Technical Report DISCS, National University of Singapore, Singapore, 1993.

[20] Martinez, T.R. and Campbell, D.M., A self-adjusting dynamic logic module, *J. Parallel Distributed Process.*, 11, 303, 1991.

[21] Minsky, M.L. and Papert, S.A., *Perceptrons*, expanded edition, MIT Press, Cambridge, 1988.

[22] Mozer, M.C. and Smolenski, P., Skeletonization: a technique for trimming the fat from a network via relevance assessment, in *Neural Information Processing Systems*, Vol. 1, Touretzky, D., Ed. Morgan Kaufmann, San Mateo, CA, 1989.

[23] Michalski, R.S., Learning by being told and learning from examples: an experimental comparison of the two methods of knowledge acquisition in the context of developing an expert system for soybean disease diagnosis, *Int. J. Policy Anal. Inf. Syst.*, 4(2), 125, 1980.

[24] Oden, G.C., A symbolic superstrate for connectionist models, IEEE ICNN, San Diego, CA, July 1988.

[25] Quinlan, J.R., Discovering rules by induction from large collections of examples, in *Expert Systems in the Micro Electronic Age*, Michie, D., Ed., Edinburgh University Press, Edinburgh, 1979.

[26] Romaniuk, S.G. and Hall, L.O., Divide and conquer networks, *Neural Networks*, 6, 1105–1116.

[27] Romaniuk, S.G. and Hall, L.O., SC-net: a hybrid connectionist, symbolic system, *Inf. Sci.*, 71, 223, 1993.

[28] Romaniuk, S.G., Pruning divide & conquer networks, *Network*, 4, 481, 1993.

[29] Romaniuk, S.G., Evolutionary growth perceptrons, in *Genetic Algorithms: Proc. 5th Int. Conf.*, Forrest, S., Ed., Morgan Kaufmann, San Mateo, CA, 1993.

[30] Rummelhart, D.E. and McClelland, J.L., Eds., *Parallel Distributed Processing: Exploration in the Microstructure of Cognition*, Vol. 1, MIT Press, Cambridge, 1986.

[31] Sanger, T.D., A tree-structured adaptive network for function approximation in high dimensional space, *IEEE Trans. Neural Networks*, 2(2), 285, 1991.

[32] Smotroff, I.G., Friedman, D.H., and Connolly, D., Self organizing modular neural networks, in *Int. Joint Conf. on Neural Networks*, Seattle, WA, 1991, II-187.

[33] Shastri, L., *Semantic Networks: An Evidential Formalization and its Connectionist Realization*, Morgan Kaufmann, San Mateo, CA, 1988.

[34] Towell, C.G. and Shavlik, J.W., Directed Propagation of Training Signals through Knowledge-Based Neural Networks, Technical Report #989, Department of CS, University of Wisconsin, Madison, 1990.

[35] Wasserman, P.D., *Neural Computing, Theory and Practice*, Van Nostrand Rheinold, New York, 1989.

[36] Valiant, L.G., A theory of the learnable, *Comm. Assoc. Comput. Mach.*, 27(11), 1134, 1984.

10

Genetic Programming of Logic-Based Neural Networks

Vincent Charles Gaudet

Abstract   Genetic algorithms and genetic programming are genetics-based optimization methods in which potential solutions evolve via operators such as selection, crossover, and mutation. Logic-based neural networks are a variation of standard neural networks; they fill the gap between distributed, unstructured neural networks and symbolic programming. In this chapter, the "Genetic Programming Paradigm" is modified in order to obtain logic-based neural networks. Modifications introduced in the chapter include connection weights on the parse trees, a new mutation operator, a new crossover operator, and a new method for generating the individuals in the initial population. The algorithm is meant to be part of a two-level development process where, at first, satisfactory logic-based neural networks are obtained using our algorithm; then, gradient-based learning methods are used to refine the networks. Satisfactory results are obtained for a 6-input logic-based neural network problem.

10.1   Introduction

Genetic algorithms [2] and genetic programming [4] are approximate optimization methods; populations of potential solutions to a problem evolve from generation to generation through genetics-based operators such as selection, crossover, and mutation.

Genetic algorithms tackle problems which are more numerical in nature, such as determining weights for standard artificial neural network, whereas Koza's "genetic programming paradigm" (GPP) [4] tackles more symbolic-oriented problems such as Boolean function learning and game-playing strategy learning.

Logic-based neural networks (LNN), defined in Reference [5], arise from the gap between well-structured symbolic programming methods and less structured, more distributed neural networks. While the lack of rigid structure in neural networks permits incredible learning capabilities at the numerical level, it also makes it harder for the user to understand what is going on within the network at the logical level. In symbolic programming, on the other hand, knowledge is very explicitly represented, and the learning capabilities are not great. LNNs have the general structure of neural networks, with neurons and weighted connections, but the neurons are based on fuzzy-set theory, thus incorporating a more symbolic structure within the network.

In this chapter we introduce modifications to the GPP that make it possible for the GPP to obtain LNN. The modifications include a new, more general data structure which is very suitable to LNNs; also, new operators that are suitable to LNNs are developed.

The LNNs obtained using the GPP method are starting architectures from which more refined, gradient-based learning is performed using LNN-adapted learning methods. Thus, for our version of the GPP, we are concentrating on two-level semiparametric development of LNN, rather than on one-level parametric learning.

Note that the class of problems tackled here by our version of the GPP is much more specific than tackled by Koza's original version. In particular, we have developed an algorithm whose data structure can include weighted connections; also, the problems tackled by our algorithm have a more layered structure.

The material is organized into several sections. We proceed with a concise description of genetic programming, contrasting that with standard genetic algorithms.

10.2   Description of Genetic Programming

10.2.1   Genetic Algorithms

The standard genetic algorithm (GA) is an approximate optimization method, where potential optimal solutions are coded in the form of bit strings [2].

The GA has a "population" of strings on which in operates to create one generation of the population from the previous, in an attempt to improve on performance and tend toward an optimum value. It does this through operators called "selection," " crossover," and "mutation."

For each string in the population, there is an associated fitness value that describes how "good" that particular string is.

The selection operator determines which members of the population will be retained, intact or modified, for the next generation. This is based on fitness values; strings (or individuals) with higher fitness values are assigned higher probabilities of getting retained for the next generation, compared to those with lower fitness values. From two selected strings, the crossover operator creates two new strings by cutting the input strings at the same point and switching their tail ends; again, fitness values affect the probability of being selected for this procedure. The cut-point is usually chosen randomly. Finally, the mutation operator changes the value of a randomly chosen bit in a selected bit string, whose probability of being chosen is, as always, dependent on fitness value.

10.2.2 Genetic Programming: Data Structure

The GPP uses the general ideas found in the standard GA, but operates on a different data structure: trees replace bit strings; each tree represents the parse tree for a given program.

For example, the tree in Figure 10.1 represents the Boolean function

$$F = (A \text{ } AND \text{ } B) \text{ } OR \text{ } [(\text{ } NOT \text{ } A) \text{ } AND \text{ } (NOT \text{ } B)].$$

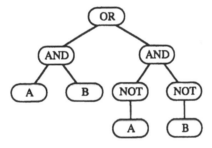

FIGURE 10.1
Example of parse tree for genetic programming.

Note that many different trees can represent the same given function. Observe also that there are no numerical values associated with the connections.

10.2.3 Crossover

This tree data structure, being more complex yet sophisticated than bit strings, requires new crossover and mutation operators.

In a standard bit-string-driven GA, crossover is performed by selecting two strings, choosing a "crossover point," and cutting and splicing the two strings at that same point. An identical concept of "crossover point" is not applicable to trees, since it would nearly imply that the two input trees in question are, from the root to the cut-point, isomorphic, an excessively restraining hypothesis. This is addressed by choosing a cut-point for each of the input trees and switching subtrees, as illustrated in Figures 10.2 and 10.3.

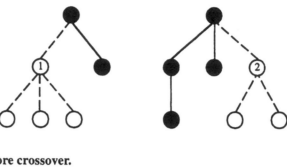

FIGURE 10.2
Two trees before crossover.

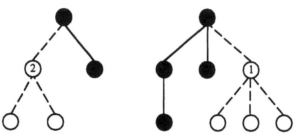

FIGURE 10.3
Two trees after crossover.

Figure 10.2 shows the two input trees, with selected crossover points. The crossover points are the two nodes, numbered "1" and "2"; the subtrees being switched are shown in white and with dashed connections. Note that the connection from the crossover point to its parent node is included with the subtree; this will become important later. For the problems tackled by the generic GPP, the connections are not weighted, and thus it is not necessary, in that case, to include the connection to the parent node. Also, to make things consistent, the root node is connected "to nowhere" via an "invisible" connection that has no effect on the functioning of tree; if the root node is chosen as crossover point, this invisible connection becomes the connection to the tree's new parent node, and the other subtree's "real" connection becomes invisible.

10.2.4 Mutation

The mutation operator in a GA flips a randomly selected bit within the bit string. In a certain way, one can think of this as randomly assigning a different value to the selected bit; there being only one such thing, we are not faced with the question of what we are mutating to.

One could create an operator that simply mimics the GA's mutation operator and changes the nature of a randomly selected node (e.g., change from an AND to an OR node). However, this would be infeasible in many cases for the simple fact that two given nodes will very often have different arities, thus resulting in conflicts for the children of the node.

The GPP's mutation operator solves this problem. It randomly picks a mutation point (a node in the tree); it then destroys the subtree (along with the connection to its parent) whose root is the mutation point, and replaces it with a new randomly generated subtree with its own connection to the parent node.

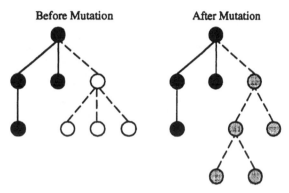

FIGURE 10.4
Tree before and after mutation.

In Figure 10.4, the tree on the left has been selected for mutation; the subtree that will get mutated is dashed and white. The tree after mutation is shown on the right; the new subtree is dashed and gray.

10.3 Description of Logic-Based Neural Networks

LNN's are explained in detail in Reference [5]. Rather than having one single standardized type of neuron, as in standard artificial neural networks, LNN's incorporate the idea of having neurons which have responses based on fuzzy set theory. The responses are similar to those of the AND and OR functions in Boolean logic.

AND neurons are defined by AND: $[0, 1] \times [0, 1] \rightarrow [0, 1]$, with response A AND B $\equiv$ min(A,B). For more than two inputs, AND is defined associatively. Figure 10.5 presents a table, for inputs restricted to $\{0, 1\} \times \{0, 1\}$, showing that the AND neuron is a generalization of the AND function encountered in Boolean logic.

Similarly, OR neurons have the response A OR B $\equiv$ max(A, B) and are defined associatively for more than two inputs. Once more we can see in Figure 10.6 that the response generalizes Boolean logic.

The NOT function is defined by NOT A $\equiv 1-$ A. Once more, this is consistent with Boolean logic, though we should state clearly that we are defining NOT as a

A	B	min(A, B)
0	0	0
0	1	0
1	0	0
1	1	1

FIGURE 10.5
2-Input AND function.

A	B	max(A, B)
0	0	0
0	1	1
1	0	1
1	1	1

FIGURE 10.6
2-Input OR function

function and will not use it as a specific node; rather, as we will see below, NOT will be used implicitly within connections.

As in neural networks, connections for LNN's are weighted:

1. If the parent of a connection is an AND neuron, the input to the neuron is
   <parent neuron input> = *<child node output>* OR *<connection weight>*

2. If the parent is an OR neuron, the input to the neuron is
   <parent neuron input> = *<child node output>* AND *<connection weight>*

Weights always have values in the interval [0, 1]; however, they may also be "inhibitive"; with an inhibitive weight, we apply the required AND or OR weightage and then apply the NOT operation before the result is sent to the parent neuron.

For example, if the parent neuron is an AND function and the weight is inhibitive, then the neuron input would be the following:

<parent neuron input> = NOT [*<child node output>* OR *<connection weight>*]

10.4   Description of the GPP for Logic-Based Neural Networks

There are four basic differences between this algorithm and Koza's GPP: our version introduces connection weights, a new mutation operator, a new crossover operator, and a new method of generating the individuals in the initial random population.

10.4.1   Coding

Neurons in the LNN are coded as nodes in Koza's parse trees. The function set (set of all nodes with arity $\geq$ 1) is

$$F = \{AND, OR\}$$

where we use arities of 2 and 3 for both the AND and OR functions. The terminal set (set of all possible nodes with arity $= 0$) is

$$T = \{\text{Set of inputs to the given problem}\}$$

The generic version of the GPP does not use weighted connections. They are not necessary in the problems it tackles. However, since we are dealing here with problems whose solutions have a weighted structure, we find it natural and also necessary to introduce weights to the connections between neurons. The weights are coded with the weight from a node to its parent being coded alongside the node.

In order to control the eventual mutation of weights, we will assign weights from symbolic categories named small (S), medium (M), or large (L); also, we determine whether the weight is normal or inhibitive. Then, when determining fitness values, we will assign a numerical value to each symbolic value; this numerical value will be the actual weight.

There are many ways to assign numerical weight values. All are probabilistic in nature. As an economy measure, we have decided to use uniform distributions on user predefined intervals.

S:	$[0, s]$
M:	$[m_1, m_2]$
L:	$[l, 1]$

where there may or may not be overlap between intervals. Also, we assign equal probabilities to normal and inhibitive weights.

10.4.2   Genetic Mechanisms

The generic GPP mutation operator destroys a randomly chosen subtree and replaces it with a newly generated subtree.

The generic GPP crossover operator picks a random subtree in each of two input trees and switches them. This crossover, alone, is sufficient to attain genetic diversity because the two crossover points are not necessarily at the same point in the two trees and switching subtrees from different locations can and will introduce new nodes to locations that previously had never seen such types of nodes. (Crossover in a standard binary-string-driven GA is not sufficient to attain genetic diversity because the two crossover points must be the same in order to maintain string length.) Thus, the generic crossover operator in GPP is sufficient to attain genetic diversity, without requiring the generic mutation operator.

Also, the generic mutation operator is undesirable since it randomly destroys information. A more useful version of this particular mutation operator would take into account the "sub-fitness" of a subtree before actually destroying it.

However, a mutation operator is required to attain genetic diversity for connection weights.

This is at the heart of the mutation operator we have defined and will be using instead of the generic mutation operator: our new mutation operator mutates connection weights. The operator picks a connection in the tree and replaces it with an "adjacent" weight; adjacency is determined as shown in Figure 10.7, where an arrow denotes adjacency between two weights. Inhibitive weights are represented by the weight symbol with an overstrike. For example, if the old connection weight is M, the new connection weight will be either S or L.

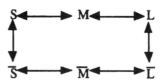

FIGURE 10.7
Adjacency between weights.

Other adjacency configurations such as the one depicted in Figure 10.8 could potentially be studied.

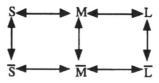

FIGURE 10.8
Other potential adjacency configuration.

Next, the standard method of generating random individuals for the initial population is far too general, especially for Boolean problems. Boolean expressions can be rewritten as a sum-of-products (SOP) or a product-of-sums (POS); knowledge of this makes it unnecessary and also wasteful to use the most general trees possible.

For a Boolean function represented as a SOP, we have an input layer followed by NOT, AND, and OR layers, the OR layer representing outputs; each layer is composed of one type of function only; there is a similar result for POS.

LNNs, which incorporate certain properties of Boolean expressions, also tend toward a similar layered structure. It is therefore natural to consider, in LNN, only trees with a tendency toward such a layered structure, each layer consisting predominantly of one type of neuron.

We incorporate a probability of nodes getting chosen at a given level. For example, we may require the second level to consist of 80% OR nodes and 20% terminals (inputs); to accomplish this, we would set the probabilities for the OR nodes to total 0.8 and the probabilities for the terminals to total 0.2. When the algorithm creates an individual, it looks at which level it is creating and chooses a node at random, with more "preferable" nodes having a higher probability of getting chosen. This tends to create individuals whose structure is pertinent to the problem.

It then follows that we should create a crossover operator that takes advantage of this layered structure. Other crossover operators have been proposed [1]; these operators limit the crossover points to two "identical" points within the trees or two points that are within the "same" subtree. However, our new crossover operator picks two crossover points that are at the same level in the tree. Thus the individuals resulting from our new crossover will have the same desired, layered structure as the individuals did before crossover. This new crossover is more general than the crossover proposed in Reference [1], but it accomplishes the same effect of preserving the underlying structure of the individuals.

10.4.3   Fitness

When determining fitness, numerical values of S, M, and L are assigned for each particular trial, as described in Section 10.4.1. Raw fitness for an individual is naturally defined as the sum of absolute values of deviations between observed and desired outputs, calculated over all possible inputs. These are averaged over the collection of trials to produce the individual's actual raw fitness; then a transformation of the form

$$\text{fitness} = \frac{1}{\varepsilon + (\text{average raw fitness})}$$

is applied, for a small $\varepsilon > 0$; this fitness measures serves a computational purpose only, when selecting individuals for a new population. The average raw fitness is a quantity really pertinent to the user, in some sense. Raw fitness varies between 0 and 2^n, though it is actually never 2^n in practice; a value of 0 signifies perfection.

A different measure of performance is the number of hits for a particular individual. A hit occurs when a particular combination of inputs produces an output within a user-specified threshold of the desired output. The numbers of hits over all combinations of inputs are averaged over the number of trials, and then rounded to the nearest integer. The hits are essentially a measure of how many outputs, on average, are acceptable to the user. Hits vary between 0 and 2^n; a value of 2^n signifies perfection.

Both these measures inform the user of the degree of approximation supplied by the individual's output, but in different ways. One might say that a good number of hits signifies an individual that currently produces the most correct outputs, whereas a good raw fitness signifies a more robust individual, capable of adapting to new weights without significant deterioration.

Also note that in the fitness function, there is no measure of the size of the network.

Algorithm	Generic Genetic Programming Paradigm	GPP For Logic-Based Neural Networks
Data Structure	General Tree	Weighted Tree
Crossover Operator	Generic Crossover Operator	Generic Crossover and Same Layer Crossover Operators
Mutation Operator	Generic Mutation Operator	Mutation of Connection Weights
Generation of Individuals in Original Population	Random with General Structure	Random with Layered Structure

FIGURE 10.9
Comparison between generic GPP and GPP for LNN.

Figure 10.9 summarizes the basic differences between the various operators, for the algorithm we tested and the generic GPP.

10.5   Experimental Studies

10.5.1   Description of Tests

The GPP for LNN was tested using a data set derived from a 6-input LNN. For testing purposes, the inputs are limited to 0 and 1.

Tests were performed using three different sets of parameters; for each set of parameters, there were 12 runs. The parameters used are listed in Figures 10.10 to 10.12.

The first parameter set tests our algorithm using only standard crossover; the second tests our algorithm using a mix of 100 standard crossovers and 100 new

Parameter Set	1	2	3
Maximum Number Of Nodes In Tree	60	60	60
Population Size	300	300	300
Number Of Generations	300	300	300
Number Of Fitness Trials	4	4	4
Number Of Generic Crossovers	200	100	0
Number Of Same Level Crossovers	0	100	200
Probability Of Weight Mutation	0.05	0.05	0.05
Threshold For Hits	0.10	0.10	0.10

FIGURE 10.10
Description of three parameter sets.

S Range	[0.0, 0.2]
M Range	[0.3, 0.7]
L Range	[0.8, 1.0]

FIGURE 10.11
Description of weight intervals.

	level 0	level 1	level 2	level 3	level 4
AND-2	0.1	0.2	0.4	0.2	0.0
AND-3	0.1	0.2	0.4	0.2	0.0
OR-2	0.4	0.3	0.05	0.0	0.0
OR-3	0.4	0.3	0.05	0.0	0.0
inputs	0.0	0.0	0.1	0.6	1.0

FIGURE 10.12
Probabilities that nodes will get selected at given levels.

crossovers per generation; the third parameter set tests our algorithm using only the new crossover operator.

The general structure of the individuals is a follows: there will be an input layer, then a layer of AND nodes, and the output layer will consist of OR nodes. Note that this is only a general structure and that there may be deviations from this structure.

10.5.2 Results

The "optimal" (64 hits) solution was not obtained within the allotted number of generations for any of the 12 runs of any of the three parameter sets. For the parameter set using 200 generic crossovers, the algorithm's best solution (in terms of average raw fitness) on average was 7.09. For the 100/100 mix, the algorithm's best solution on average was 6.91. For the parameter set using 200 new crossovers, the algorithm's best solution on average was 6.90. The new crossover operator, working alone or with the generic crossover operator, produces slightly better solutions.

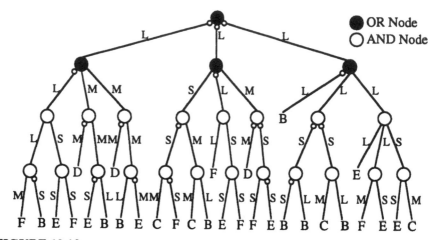

FIGURE 10.13
Best obtained logic-based neural network.

The overall best solution was obtained with the third parameter set (200 new crossover) in generation 111. It has a raw fitness of 3.82 and got 58 hits. Its tree structure is shown in Figure 10.13. Large black circles represent OR nodes, whereas large white circles represent AND nodes. Weights are indicated by their symbolic value, shown next to their respective connections; small white circles represent inhibitive weights. The inputs are at the bottom; the information travels up to the root, or output node.

Suppose that each of the 58 hits is a "true" hit where the output is identical to the desired output; then, the estimated deviation from the desired output for the 6 nonhits would be 0.64. However, a more realistic approximation of deviation from desired output would assume the hits having a deviation of approximately half the threshold value; then, the estimated deviation from the desired output for the nonhits would be reduced to 0.15. This is most certainly acceptable, since now, more refined, gradient-based learning techniques will be used in order to complete the learning process.

10.6 Conclusions

We have presented a genetic programming-based algorithm that obtains satisfactory solutions to medium-sized LNN problems. In developing this algorithm, we have introduced new operators that take advantage of knowledge we have about logic-based neural networks. We have also introduced a new weighted tree structure. Some of the methods introduced, such as the method of generating individuals in the initial random population and the new crossover operator, could possibly be applied to other types of problems where structured, layered solutions are desired.

The algorithm we have presented is meant to be part of a two-level development process. First, satisfactory LNNs are obtained using our algorithm; these networks are then refined using gradient-based learning techniques.

Some problems still need to be studied.

First, a data structure incorporating many outputs for one given network should be developed. Currently, what can be done is to have multiple runs producing results for each of the separate outputs. This, however, does not produce optimal solutions, since, very often, portions of networks may be combined.

Second, testing has to be performed using more general "continuous-to-continuous" intervals rather than the "discrete-to-continuous" intervals used so far.

Acknowledgments

Support from the Natural Sciences and Engineering Research Council and MICRONET is gratefully acknowledged.

Also, I would like to thank Professor Witold Pedrycz for the numerous hours spent discussing genetic programming and neural networks and for the advice and assistance he has given me during my years of collaboration with him.

References

[1] D'haeseleer, P., Context Preserving Crossover in Genetic Programming, 1994 (also available via anonymous ftp from ftp.cc.utexas.edu under pub/genetic programming/papers with name WCCI94_CPC.ps.Z).

[2] Goldberg, D.E., *Genetic Algorithm in Search, Optimization, and Machine Learning*, Addison-Wesley; Reading, MA, 1989.

[3] Keith, M. and Martin, M., Genetic programming in C++: implementation issues, in *Advances in Genetic Programming*, MIT Press, Cambridge, 1994, 285.

[4] Koza, J.R., *Genetic Programming*, MIT Press, Cambridge, 1992, 1.

[5] Pedrycz, W. and Rocha, A.F., Fuzzy-set based models of neurons and knowledge-based networks, *IEEE Trans. Fuzzy Syst.*, 1(4), 254, 1993.

11

Construction of Fuzzy Classification Systems with Linguistic If–Then Rules Using Genetic Algorithms

Hisao Ishibuchi, Tadahiko Murata, and Hideo Tanaka

Abstract This chapter proposes a genetic-algorithm-based method for constructing a fuzzy classification system with linguistic if–then rules. In our method, first a large number of linguistic rules are generated from numerical data, *i.e.*, from training patterns. Then only a small number of significant rules are selected by a genetic algorithm in order to construct a compact fuzzy classification system. A set of linguistic rules is coded as an individual in the genetic algorithm. The objectives of our rule selection problem are to maximize the number of correctly classified training patterns and to minimize the number of selected linguistic rules. The fitness value of each individual is defined by these objectives. The proposed method is illustrated by computer simulations on a numerical example and the well-known iris data. We also propose a hybrid approach that incorporates a learning procedure into the genetic algorithm. The grade of certainty of each linguistic rule in each individual is adjusted by the learning procedure during the execution of the genetic algorithm. It is demonstrated by computer simulations on the iris data that the hybrid algorithm can find a small number of linguistic rules with high classification power.

11.1 Introduction

Fuzzy logic with fuzzy if–then rules has been mainly applied to control problems [1, 2]. In most fuzzy control systems, fuzzy rules were derived and tuned

227

by human experts. Since the generation of fuzzy rules and their tuning is time-consuming for human experts, several approaches have been proposed for automatically generating fuzzy rules from numerical data (for example, see References [3] to [5]). Self-learning methods have also been proposed for adjusting membership functions of fuzzy sets in fuzzy rules. For example, Ichihashi and Watanabe [6] and Nomura et al. [7] proposed gradient descent methods for the learning of fuzzy rules. Takagi and Hayashi [8], Lin and Lee [9], Jang [10], and Horikawa et al. [11] proposed neural-network-based methods for the generation of fuzzy rules and their tuning.

Genetic algorithms [12, 13] have also been employed for generating fuzzy rules and adjusting membership functions of fuzzy sets. For example, membership functions were adjusted by genetic algorithms in Karr [14] and Karr and Gentry [15]. Fuzzy partitions of input spaces were determined in Nomura et al. [16] and Fukuda et al. [17]. That is, both the number of fuzzy sets and the membership function of each fuzzy set were determined. In Thrift [18], an appropriate fuzzy set in the consequent part of each fuzzy rule was selected. These approaches applied genetic algorithms to fuzzy control problems by coding a fuzzy rule table (*i.e.*, a set of fuzzy rules) as an individual. On the other hand, Valenzuela-Rendon [19] proposed a fuzzy classifier system where a single fuzzy rule was coded as an individual. Appropriate fuzzy sets in the antecedent and consequent parts of each fuzzy rule were selected by the fuzzy classifier system in Reference [19].

While various methods have been proposed for generating fuzzy rules and adjusting membership functions, only a few approaches have dealt with classification problems. Ishibuchi et al. [20, 21] proposed a generation method of fuzzy rules from numerical data for classification problems. Genetic-algorithm-based methods for selecting fuzzy rules were proposed in Ishibuchi et al. [22, 23] where a small number of fuzzy rules were selected from a large number of candidate rules by genetic algorithms.

In this chapter, we propose a genetic-algorithm-based approach to the construction of a fuzzy classification system with linguistic if–then rules. A small number of linguistic rules are selected by a genetic algorithm to construct a compact fuzzy classification system. The main advantage of the proposed approach over our former work [22, 23] is the clarity of the selected rules. That is, human decision makers can easily understand each of the selected rules because they are linguistic rules. The proposed method is illustrated by computer simulations on a numerical example and the well-known iris data (see Fisher [24]). A hybrid approach that incorporates a learning procedure into the genetic algorithm is also proposed in order to improve the performance of fuzzy classification systems. The grade of certainty of each linguistic rule is adjusted by the learning procedure during the execution of the genetic algorithm. It is demonstrated by computer simulations on the iris data that the hybrid algorithm can find a small number of linguistic rules with high classification power.

11.2   Automatic Generation of Linguistic If–Then Rules

11.2.1   Rule Generation

For simplicity, let us consider a two-class classification problem in the two-dimensional pattern space $[0, 1]^2$. We assume that m patterns, $x_p = (x_{p1}, x_{p2})$, $p = 1, 2, \cdots, m$, are given as training patterns. Each pattern belongs to one of the two classes: class 1 (C1) and class 2 (C2). These patterns are used for generating linguistic if–then rules.

When we generate linguistic if–then rules, first we should specify linguistic labels employed in the antecedent part of each rule. In this chapter, we use the five labels: *small, medium small, medium, medium large,* and *large.* The meaning of each linguistic label is specified by its membership function. The membership functions of the five linguistic labels are shown in Figure 11.1. These membership functions are written as follows:

$$\mu_{small}(x) = \max\{0, 1 - 4 \cdot |x - 0|\}, \tag{11.1}$$

$$\mu_{medium\ small}(x) = \max\{0, 1 - 4 \cdot |x - 0.25|\}, \tag{11.2}$$

$$\mu_{medium}(x) = \max\{0, 1 - 4 \cdot |x - 0.5|\}, \tag{11.3}$$

$$\mu_{medium\ large}(x) = \max\{0, 1 - 4 \cdot |x - 0.75|\}, \tag{11.4}$$

$$\mu_{large}(x) = \max\{0, 1 - 4 \cdot |x - 1|\}, \tag{11.5}$$

where x is a real number in the unit interval $[0,1]$.

Let us consider a two-class classification problem in the two-dimensional pattern space $[1, 0]^2$ shown in Figure 11.2. We can intuitively generate the following linguistic if–then rules from the given patterns.

If x_{p1} is *small* and x_{p2} is *small* then (x_{p1}, x_{p2}) belongs to Class 1,
If x_{p1} is *small* and x_{p2} is *medium small* then (x_{p1}, x_{p2}) belongs to Class 1,
If x_{p1} is *small* and x_{p2} is *medium* then (x_{p1}, x_{p2}) belongs to Class 1,
$\cdots$
If x_{p1} is *large* and x_{p2} is *medium large* then (x_{p1}, x_{p2}) belongs to Class 1,
If x_{p1} is *large* and x_{p2} is *large* then (x_{p1}, x_{p2}) belongs to Class 2.

These linguistic rules are written in a rule table in Figure 11.3. While we intuitively derived the linguistic rules in Figure 11.3 from the given patterns in Figure 11.2,

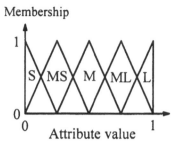

FIGURE 11.1
Membership functions of five linguistic labels (S: *small*, MS: *medium small*, M: *medium*, ML: *medium large*, L: *large*).

such intuitive rule generation is not always easy, especially for classification problems in high dimensional pattern spaces.

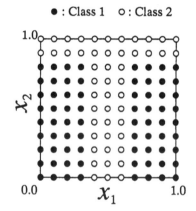

FIGURE 11.2
A classification problem.

In order to specify the relative importance of each rule, we attach the grade of certainty to each rule. For example, we use the following linguistic rules with the grade of certainty:

If x_{p1} is *small* and x_{p2} is *small*
     then (x_{p1}, x_{p2}) belongs to Class 1 with $CF = 1.00$,
If x_{p1} is *small* and x_{p2} is *medium small*
     then (x_{p1}, x_{p2}) belongs to Class 1 with $CF = 1.00$,
    . . .

If x_{p1} is *large* and x_{p2} is *large*
     then (x_{p1}, x_{p2}) belongs to Class 2 with $CF = 0.78$.

---

where CF is the grade of certainty of each rule.

x_2 \ x_1	S	MS	M	ML	L
L	C2	C2	C2	C2	C2
ML	C1	C1	C2	C1	C1
M	C1	C1	C2	C1	C1
MS	C1	C1	C2	C1	C1
S	C1	C1	C2	C1	C1

C1:Class 1; C2: Class 2

FIGURE 11.3
The rule table with the intuitively generated 25 linguistic rules.

More generally, we use the following linguistic rules when we have K linguistic labels $A_1, A_2, \cdots, A_K$ for each axis of the pattern space:

$$\text{Rule } R_{ij} : \text{ If } x_{p1} \text{ is } A_i \text{ and } x_{p2} \text{ is } A_j$$
$$\text{then } (x_{p1}, x_{p2}) \text{ belongs to Class } C_{ij} \text{ with } CF = CF_{ij}, \qquad (11.6)$$
$$i = 1, 2, \cdots, K; j = 1, 2, \cdots, K,$$

where R_{ij} is the label of the linguistic rule, A_i and A_j are linguistic labels, C_{ij} is the consequent class, and CF_{ij} is the grade of certainty. In the same manner as in Ishibuchi et al. [20] to [23], the consequent class C_{ij} and the grade of certainty CF_{ij} of each linguistic rule in Equation 11.6 can be automatically determined from the given training patterns by the following procedure:

Procedure 1: generation procedure of linguistic rules

Step 1: Calculate β_{C1} and β_{C2} as

$$\beta_{C1} = \sum_{x_p \in C1} \mu_i(x_{p1}) \cdot \mu_j(x_{p2}), \qquad (11.7)$$

$$\beta_{C2} = \sum_{x_p \in C2} \mu_i(x_{p1}) \cdot \mu_j(x_{p2}), \qquad (11.8)$$

where $\mu_i(x_{p1})$ and $\mu_j(x_{p2})$ are the membership functions of the linguistic labels A_i and A_j, respectively. In Equations 11.7 and 11.8, β_{C1} and β_{C2}

can be viewed as indexes that measure the number of compatible patterns with the antecedent of the linguistic rule R_{ij}.

Step 2: Determine the consequent class C_{ij} as

$$C_{ij} = \begin{cases} C1, & \text{if } \beta_{C1} > \beta_{C2}, \\ C2, & \text{if } \beta_{C1} < \beta_{C2}, \\ \phi, & \text{if } \beta_{C1} = \beta_{C2}. \end{cases} \tag{11.9}$$

When $\beta_{C1} = \beta_{C2}$, we cannot specify the consequent class C_{ij}. Therefore we assign ϕ to C_{ij} if $\beta_{C1} = \beta_{C2}$. When no patterns are compatible with the antecedent of the linguistic rule R_{ij} (i.e., when $\beta_{C1} = \beta_{C2} = 0$), ϕ is also assigned to C_{ij}.

Step 3: The grade of certainty is determined as

$$CF_{ij} = \begin{cases} (\beta_{C1} - \beta_{C2})/(\beta_{C1} + \beta_{C2}), & \text{if } \beta_{C1} > \beta_{C2}, \\ (\beta_{C2} - \beta_{C1})/(\beta_{C1} + \beta_{C2}), & \text{if } \beta_{C1} < \beta_{C2}, \\ 0, & \text{if } \beta_{C1} = \beta_{C2}. \end{cases} \tag{11.10}$$

Linguistic rules with ϕ in the consequent part are dummy rules that have no effect in the classification phase of new patterns. If there are no patterns that are compatible with the antecedent of the linguistic rule R_{ij}, a dummy rule is generated.

The grade of certainty CF_{ij} specified by Equation 11.10 has the following intuitively acceptable two properties:

1. The grade of certainty CF_{ij} assumes its maximum value (i.e., $CF_{ij} = 1$) when all the patterns that are compatible with the linguistic rule R_{ij} belong to the same class. This situation is illustrated in Figure 11.4 where there is no Class 2 pattern that is compatible with the rule R_{ij}. Therefore, we have the following rule for the case of Figure 11.4.
    Rule R_{ij}: If x_{p1} is A_i and x_{p2} is A_j
            then (x_{p1}, x_{p2}) belongs to Class 1 with $CF = 1$.

2. The grade of certainty CF_{ij} assumes its minimum value (i.e., $CF_{ij} = 0$) when the values of β_{C1} and β_{C2} are the same. This situation is illustrated in Figure 11.5 where the grades of compatibility of the four compatible patterns are the same [i.e., $\mu_i(x_{p1}) \cdot \mu_j(x_{p2}) = 0.25$ for those patterns]. Therefore we have the following rule for the case of Figure 11.5.
    Rule R_{ij}: If x_{p1} is A_i and x_{p2} is A_j
            then (x_{p1}, x_{p2}) belongs to Class ϕ with $CF = 0$.

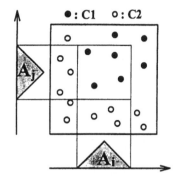

FIGURE 11.4
An example of training patterns for which the grade of certainty of the rule R_{ij} is maximum.

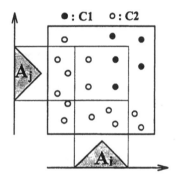

FIGURE 11.5
An example of training patterns for which the grade of certainty of the rule R_{ij} is minimum.

11.2.2 Classification Using Generated Rules

By the above procedure, we can generate the K^2 linguistic rules in Equation 11.6 because we have the K linguistic labels $A_1, A_2, \cdots, A_K$ for each axis of the two-dimensional pattern space. Let us denote the set of all the generated linguistic rules by S_{ALL}:

$$S_{ALL} = \{R_{ij} | i = 1, 2, \cdots, K; j = 1, 2, \cdots, K\}. \qquad (11.11)$$

We denote a subset of S_{ALL} by S. When a fuzzy classification system consists of the linguistic rules in the subset S, a new pattern $x_p = (x_{p1}, x_{p2})$ is classified by the following procedure.

Procedure 2: classification of a new pattern $x_p = (x_{p1}, x_{p2})$

Step 1: Calculate α_{C1} and α_{C2} as follows:

$$\alpha_{C1} = \max\{\mu_i(x_{p1}) \cdot \mu_j(x_{p2}) \cdot CF_{ij} | C_{ij} = C1 \text{ and } R_{ij} \in S\} \quad (11.12)$$

$$\alpha_{C2} = \max\{\mu_i(x_{p1}) \cdot \mu_j(x_{p2}) \cdot CF_{ij} | C_{ij} = C2 \text{ and } R_{ij} \in S\} \quad (11.13)$$

Step 2: If $\alpha_{C1} > \alpha_{C2}$ then classify x_p as class 1. If $\alpha_{C1} < \alpha_{C2}$ then classify x_p as class 2. If $\alpha_{C1} = \alpha_{C2}$ then the classification of x_p is rejected (*i.e.*, x_p is left as an unclassifiable pattern).

In this procedure, the inferred class is the consequent of the linguistic rule that has the maximum value of $\mu_i(x_{p1}) \cdot \mu_j(x_{p2}) \cdot CF_{ij}$ among all the rules in S.

Using the classification problem in Figure 11.2, we explain the rule generation procedure and the classification procedure. If we use the five linguistic labels in Figure 11.1, we can generate 25 linguistic rules by the rule generation procedure: Procedure 1. The generated linguistic rules are shown in Figure 11.6 where the grade of certainty of each linguistic rule is written in the parentheses under the consequent class. From the comparison between Figures 11.3 and 11.6, we can see that the consequent classes of the linguistic rules generated by Procedure 1 are exactly the same as those of the intuitively generated rules in Figure 11.3.

x_2 \ x_1	S	MS	M	ML	L
L	C2 (0.78)	C2 (0.81)	C2 (0.97)	C2 (0.81)	C2 (0.78)
ML	C1 (0.67)	C1 (0.39)	C2 (0.74)	C1 (0.39)	C1 (0.67)
M	C1 (1.00)	C1 (0.67)	C2 (0.69)	C1 (0.67)	C1 (1.00)
MS	C1 (1.00)	C1 (0.67)	C2 (0.69)	C1 (0.67)	C1 (1.00)
S	C1 (1.00)	C1 (0.67)	C2 (0.69)	C1 (0.67)	C1 (1.00)

C1:Class 1; C2: Class 2

FIGURE 11.6
The rule table with the 25 linguistic rules generated by Procedure 1.

When all the generated 25 linguistic rules in Figure 11.6 are used in a fuzzy classification system, the pattern space is divided into two decision areas by the classification procedure (*i.e.*, Procedure 2) as shown in Figure 11.7. We can see from Figure 11.7 that all the given patterns are correctly classified by the 25 linguistic rules in Figure 11.6.

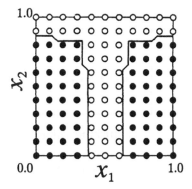

FIGURE 11.7
The boundary between the two classes drawn by Procedure 2 using the 25 linguistic rules in Figure 11.6.

11.2.3 An Extension

From the rule table in Figure 11.6, we can see that the consequent class is always $C2$ if x_1 is *medium* or x_2 is *large*. This motivates us to use the following two linguistic rules for the classification problem in Figure 11.2:

If x_{p1} is *medium* then (x_{p1}, x_{p2}) belongs to Class 2,
If x_{p2} is *large* then (x_{p1}, x_{p2}) belongs to Class 2.

In the first rule, x_{p2} can assume any value in the domain of x_2 (*i.e.*, in the unit interval [0, 1]). Therefore x_2 is a "don't care attribute" in this rule. On the other hand, x_1 is a "don't care attribute" in the second rule.

In order to handle "don't care attributes" in the framework of our linguistic rules, we regard the term *"don't care"* as a new linguistic label that has the following membership function:

$$\mu_{don't\ care}(x) = \begin{cases} 1, & if\ 0 \le x \le 1 \\ 0, & otherwise \end{cases} \tag{11.14}$$

This membership function is depicted in Figure 11.8.

By regarding *"don't care"* as a new linguistic label with the membership function in Equation 11.14, we can determine the grade of certainty of each of the above two linguistic rules in the same manner as in the other linguistic rules in Figure 11.6. By the rule generation procedure (*i.e.*, Procedure 1), we can calculate the grades of certainty of the above two rules as follows:

If x_{p1} is *medium* and x_{p2} is *don't care*
   then (x_{p1}, x_{p2}) belongs to class 2 with $CF = 0.75$,
If x_{p1} is *don't care* and x_{p2} is *large*
   then (x_{p1}, x_{p2}) belongs to class 2 with $CF = 0.84$.

Membership

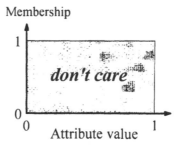

FIGURE 11.8
The membership function of the linguistic label *"don't care."*

Since any combination of *"don't care"* and the other linguistic labels is possible, we have the rule table in Figure 11.9 where DC means *"don't care."* It should be noted that the rule table in Figure 11.9 includes all the linguistic rules in Figure 11.6. Because the generation of a linguistic rule by Procedure 1 is independent of other rules, the same linguistic rules are generated in Figure 11.9 as in Figure 11.6 where the linguistic label *"don't care"* was not taken into account.

x_2 \ x_1	DC	S	MS	M	ML	L
L	C2 (0.84)	C2 (0.78)	C2 (0.81)	C2 (0.97)	C2 (0.81)	C2 (0.78)
ML	C1 (0.21)	C1 (0.67)	C1 (0.39)	C2 (0.74)	C1 (0.39)	C1 (0.67)
M	C1 (0.45)	C1 (1.00)	C1 (0.67)	C2 (0.69)	C1 (0.67)	C1 (1.00)
MS	C1 (0.45)	C1 (1.00)	C1 (0.67)	C2 (0.69)	C1 (0.67)	C1 (1.00)
S	C1 (0.45)	C1 (1.00)	C1 (0.67)	C2 (0.69)	C1 (0.67)	C1 (1.00)
DC	C1 (0.19)	C1 (0.64)	C1 (0.36)	C2 (0.75)	C1 (0.36)	C1 (0.64)

C1:Class 1;   C2: Class 2

FIGURE 11.9
The rule table with the 36 linguistic rules generated by Procedure 1.

By regarding *"don't care"* as a linguistic label, we generated the 36 linguistic rules in Figure 11.9. Since some linguistic rules overlap each other, all the generated 36 linguistic rules are not necessary in the classification phase. If a small number of necessary linguistic rules are selected from the generated rules, we can construct a compact fuzzy classification system. Then we can easily examine each of the selected linguistic rules to understand the constructed classification system.

Such a careful rule examination is practically impossible if there are hundreds of linguistic rules in the constructed classification system ($6^3 = 216$ rules can be generated for three-dimensional problems, and $6^4 = 1296$ rules for four-dimensional problems). In the next section, we show how a small number of linguistic rules can be selected from a large number of candidate rules by a genetic algorithm.

11.3 Application of Genetic Algorithms

11.3.1 Formulation of a Combinatorial Optimization Problem

Using the rule generation procedure in Section 11.2, the following linguistic rules were generated:

$$\text{Rule } R_{ij} : \text{If } x_{p1} \text{ is } A_i \text{ and } x_{p2} \text{ is } A_j$$
$$\text{then } (x_{p1}, x_{p2}) \text{ belongs to class } C_{ij} \text{ with } CF = CF_{ij}, \qquad (11.15)$$
$$i = 1, 2, \cdots, K; \, j = 1, 2, \cdots, K,$$

where A_i and A_j are linguistic labels and $K = 6$.

Let us denote the set of all the generated linguistic rules by S_{ALL} and a subset of S_{ALL} by S. Our problem in this section is to find a compact rule set S with high classification power by selecting a small number of significant linguistic rules from S_{ALL}. Therefore the objectives of our rule selection problem are to maximize the number of correctly classified training patterns by S and to minimize the number of linguistic rules in S. This problem can be formulated as the following two-objective combinatorial optimization problem.

Problem 1: Maximize $NCP(S)$ and minimize $|S|$, subject to $S \subseteq S_{ALL}$, where $NCP(S)$ is the number of correctly classified training patterns by S and $|S|$ is the cardinality of S (*i.e.*, the number of linguistic rules in S).

Since Problem 1 has two objectives, it is difficult to directly apply a genetic algorithm to this problem. Thus we modify Problem 1 by introducing positive weights W_{NCP} and W_S as follows.

Problem 2: Maximize $W_{NCP} \cdot NCP(S) - W_S \cdot |S|$ subject to $S \subseteq S_{ALL}$. In general, the classification power of a classification system is more important than its compactness. Therefore the weights in Problem 2 should be specified as $0 < W_S \ll W_{NCP}$.

11.3.2   Formulation of a Fitness Function

In Ishibuchi et al. [22], we employed the objective function of Problem 2 as a fitness function in a genetic algorithm for selecting fuzzy if–then rules. From various computer simulations on several classification problems, we found that a simple modification of the objective function led to better performance of genetic algorithms. Therefore we use the modified objective function as a fitness function in this chapter.

The modification is based on an intuitive idea that a linguistic rule compatible with a large pattern subspace can classify many patterns. For example, if we compare the following two linguistic rules, the first rule can classify much more patterns than the second rule because the former is compatible with a larger pattern subspace.

If x_{p1} is *medium* and x_{p2} is *don't care*
    then (x_{p1}, x_{p2}) belongs to Class 2 with $CF = 0.5$,
If x_{p1} is *medium* and x_{p2} is *medium*
    then (x_{p1}, x_{p2}) belongs to Class 2 with $CF = 0.5$.

In order to select general linguistic rules that are compatible with large pattern subspaces, we assign small penalties to such linguistic rules and large penalties to local linguistic rules that are compatible with small pattern subspaces. First let us define the fineness of each linguistic label as follows:

$$Fineness(small) = \cdots = Fineness(large) = 5, \qquad (11.16)$$

$$Fineness(don't\ care) = 1. \qquad (11.17)$$

The fineness can be viewed as an index of the specificity of each linguistic label. The fineness of each linguistic rule in Equation 11.15 is defined as follows:

$$Fineness(R_{ij}) = Fineness(A_i) + Fineness(A_j). \qquad (11.18)$$

where $Fineness(A_i)$ and $Fineness(A_j)$ are defined in Equations 11.16 and 11.17. For example, the fineness of the first linguistic rule in the above (*i.e.*, $A_i = medium$ and $A_j = don't\ care$) is 6, and that of the second rule (*i.e.*, $A_i = medium$ and $A_j = medium$) is 10. Since the fineness is large for local rules and small for general rules, it can be used as the penalty of each linguistic rule. Using the fineness of each linguistic rule as its penalty, we define a fitness function by modifying the objective function of Problem 2 as follows:

$$f(S) = W_{NCP} \cdot NCP(S) - W_S \cdot \sum_{R_{ij} \in S} Fineness(R_{ij}). \qquad (11.19)$$

11.3.3   Coding of Each Individual

In genetic algorithms, each feasible solution of Problem 2 is treated as an individual. That is, a rule set S should be represented by a string. Let us denote a rule set S by a string $s_1 s_2 \cdots s_N$ as $S = s_1 s_2 \cdots s_N$ where

---

$N = K^2$ (*i.e.*, N is the total number of rules in S_{ALL}),
$s_r = 1$ denotes that the rth rule belongs to S,
$s_r = -1$ denotes that the rth rule does not belong to S,
$s_r = 0$ denotes that the rth rule is a dummy rule.

The index r of the linguistic rule R_{ij} is determined as

$$r = K \cdot (i - 1) + j, \tag{11.20}$$

Since dummy rules have no effect on fuzzy inference in the classification phase (*i.e.*, on Procedure 2 in Section 11.2), they should be excluded from a rule set S. Therefore they are represented as $s_r = 0$ in this coding in order to prevent S from including them.

A string $s_1 s_2 \cdots s_N$ is decoded as

$$S = \left\{ R_{ij} | s_r = 1; r = 1, 2, \ldots, N \right\}. \tag{11.21}$$

The performance of the rule set S is measured by the fitness function in Equation 11.19.

11.3.4   Genetic Operations

The following genetic operations are employed to generate and handle a set of strings (*i.e.*, a population) in our genetic algorithm for the rule selection problem.

Step 0 (Initialization):   Generate an initial population containing N_{pop} strings where N_{pop} is the number of strings in each population. In this operation, each string is generated by assigning 0 to dummy rules and randomly assigning 1 or -1 to the other rules. Each nondummy rule, which is randomly included in each string, has a 50% chance of being chosen for inclusion in each string.

Step 1 (Selection):   Select $N_{pop}/2$ pairs of strings from the current population. The selection probability $P(S)$ of a string S in a population Ψ is specified as

$$P(S) = \{f(S) - f_{\min}(\Psi)\} / \sum_{S' \in \Psi} \{f(S') - f_{\min}(\Psi)\}, \tag{11.22}$$

where

$$f_{\min}(\Psi) = \min \{f(S) | S \in \Psi\}. \tag{11.23}$$

Step 2 (Crossover):   For each selected pair, randomly choose bit positions. Each
bit position is chosen with the probability of 0.5. Interchange the bit values
at the chosen positions in the selected pair. This crossover operation is
illustrated in Figure 11.10. This type of crossover was referred to as the
uniform crossover in Syswerda [25].

Step 3 (Mutation):   For each bit value of the generated strings by the crossover
operation, apply the following mutation operation:

$s_r = 1 \rightarrow s_r = -1$ with the mutation probability $P_m(1 \rightarrow -1)$
$s_r = -1 \rightarrow s_r = 1$ with the mutation probability $P_m(-1 \rightarrow 1)$

Different mutation probabilities $P_m(1 \rightarrow -1)$ and $P_m(-1 \rightarrow 1)$ are
assigned to the mutations from 1 to -1 and from -1 to 1, respectively. A
larger probability is usually assigned to $P_m(1 \rightarrow -1)$ than to $P_m(-1 \rightarrow 1)$
in order to reduce the number of linguistic rules in each individual.

Step 4 (Elitist strategy):   Randomly remove one string from the N_{pop} strings
generated by the above operations, and add the string with the maximum
fitness value in the previous population to the current one.

Step 5 (Termination test):   If a prespecified stopping condition is not satisfied,
return to Step 1. The total number of generations is used as a stopping
condition in this chapter.

Parent 1:	s_1 s_2 s_3 s_4 s_5 s_6 s_7 ... s_N
Parent 2:	$\underline{s}_1$ $\underline{s}_2$ $\underline{s}_3$ $\underline{s}_4$ $\underline{s}_5$ $\underline{s}_6$ $\underline{s}_7$ ... $\underline{s}_N$
Positions:	* * * *
Child 1:	$\underline{s}_1$ s_2 $\underline{s}_3$ $\underline{s}_4$ s_5 $\underline{s}_6$ s_7 ... s_N
Child 2:	s_1 $\underline{s}_2$ s_3 s_4 $\underline{s}_5$ s_6 $\underline{s}_7$ ... $\underline{s}_N$

FIGURE 11.10
An example of the uniform crossover.

11.4   Simulation Results

11.4.1   Simulation Results for a Numerical Example

Let us select linguistic rules for the classification problem in Figure 11.2. By using the six linguistic labels: *small, medium small, medium, medium large, large,* and *don't care,* we had the 36 linguistic rules in Figure 11.9. Our problem is to select a small number of significant rules from those rules. The number of all possible combinations of the selected rules is $2^{36} \approx 6.9 \times 10^{10}$. While this number includes some meaningless solutions such as an empty rule set (*i.e.*, no rule is selected), 6.9×10^{10} indicates the size of our rule selection problem.

For the coding of each rule set S, the 36 linguistic rules are numbered as shown in Figure 11.11. Therefore each rule set S is represented by the string $s_1 s_2 s_3 \cdots s_{36}$.

x_2 \ x_1	DC	S	MS	M	ML	L
L	S_6	S_{12}	S_{18}	S_{24}	S_{30}	S_{36}
ML	S_5	S_{11}	S_{17}	S_{23}	S_{29}	S_{35}
M	S_4	S_{10}	S_{16}	S_{22}	S_{28}	S_{34}
MS	S_3	S_9	S_{15}	S_{21}	S_{27}	S_{33}
S	S_2	S_8	S_{14}	S_{20}	S_{26}	S_{32}
DC	S_1	S_7	S_{13}	S_{19}	S_{25}	S_{31}

FIGURE 11.11
The indexes of the 36 linguistic rules.

We applied the genetic algorithm in the last section with the following parameter specifications to our rule selection problem.

Weight values:	$W_{NCP} = 1000, W_S = 1$,
Population size:	$N_{pop} = 10$,
Crossover probability:	1.00,
Mutation probability:	$P_m(1 \rightarrow -1) = 0.1, P_m(-1 \rightarrow 1) = 0.0001$,
Stopping condition:	1000 generations.

Since there was no dummy rule in Figure 11.9, 1 or -1 was randomly assigned to each bit of each string in the initial population. An example of the string in the initial population is shown in Figure 11.12. The genetic operations described in the last section were applied to the randomly generated strings in the initial population. Then, new strings in the second population were generated. The genetic operations were iteratively applied to a current population to generate a next population. The best string in the 1000th population is shown in Figure 11.13. Since each population contained 10 strings, we can see that the string in Figure 11.13 was obtained after examining $10 \times 1000 = 10^4$ strings out of the 6.9×10^{10} possible combinations of the selected rules.

x_2 \ x_1	DC	S	MS	M	ML	L
L	1	1	1	1	-1	1
ML	-1	-1	1	1	1	-1
M	1	-1	1	-1	1	1
MS	-1	-1	1	1	-1	-1
S	1	1	-1	1	-1	1
DC	-1	-1	-1	-1	1	-1

FIGURE 11.12
A randomly generated string in the initial generation.

x_2 \ x_1	DC	S	MS	M	ML	L
L	1	-1	-1	-1	-1	-1
ML	-1	-1	-1	-1	-1	-1
M	-1	-1	-1	-1	-1	-1
MS	-1	-1	-1	-1	-1	-1
S	-1	-1	-1	-1	-1	-1
DC	1	-1	-1	1	-1	-1

FIGURE 11.13
The best string in the final generation.

By decoding the string in Figure 11.13, we have the following three linguistic rules.

If x_{p1} is *don't care* and x_{p2} is *don't care*
  then (x_{p1}, x_{p2}) belongs to Class 1 with $CF = 0.19$,
If x_{p1} is *don't care* and x_{p2} is *large*
  then (x_{p1}, x_{p2}) belongs to Class 2 with $CF = 0.84$,
If x_{p1} is *medium* and x_{p2} is *don't care*
  then (x_{p1}, x_{p2}) belongs to Class 2 with $CF = 0.75$.

We can see from the given patterns in Figure 11.2 that the selected linguistic rules coincide with our intuition very well. It should be noted that the first rule with a small grade of certainty is used for the classification of a new pattern only when the other two rules do not have high compatibility with that pattern. By the selected three rules, all the given patterns are correctly classified. The boundary between the two classes is shown in Figure 11.14.

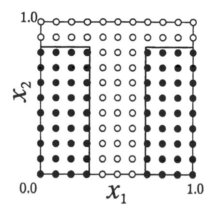

FIGURE 11.14
The boundary between the two classes drawn by Procedure 2 using the selected three rules.

We iterated the same computer simulation ten times by generating different initial populations. In all those ten trials, the same linguistic rules were selected by the genetic algorithm. This indicates the good performance of the genetic algorithm.

11.4.2   Simulation Results for Iris Data

In order to examine the performance of the proposed method for high-dimensional classification problems, we applied the genetic algorithm to the well-known iris data (see Fisher [24]). The iris data consist of the following four-dimensional patterns from three classes:

Class 1 (Iris setosa): $\boldsymbol{x}_p = (x_{p1}, x_{p2}, x_{p3}, x_{p4})$, $p = 1, 2, \ldots, 50$,
Class 2 (Iris versicolor): $\boldsymbol{x}_p = (x_{p1}, x_{p2}, x_{p3}, x_{p4})$, $p = 51, 52, \ldots, 100$,
Class 3 (Iris virginica): $\boldsymbol{x}_p = (x_{p1}, x_{p2}, x_{p3}, x_{p4})$, $p = 101, 102, \ldots, 150$,

---

where x_{p1} is the sepal length, x_{p2} is the sepal width, x_{p3} is the petal length, and x_{p4} is the petal width. In computer simulations of this section, all the attribute values were normalized into real numbers in the unit interval $[0, 1]$ as

$$x_{pi} := (x_{pi} - \min\{x_{pi}\})/(\max\{x_{pi}\} - \min\{x_{pi}\}),$$
$$p = 1, 2, \ldots, 150; \ i = 1, 2, 3, 4, \tag{11.24}$$

where

$$\min\{x_{pi}\} = \min\{x_{pi} | p = 1, 2, \ldots, 150\}, \tag{11.25}$$

$$\max\{x_{pi}\} = \max\{x_{pi} | p = 1, 2, \ldots, 150\}. \tag{11.26}$$

Therefore the classification problem of the iris data was transformed into a three-class classification problem in the four-dimensional unit cube $[0, 1]^4$.

Since the classification problem of the iris data has four attributes, we used the following rules:

Rule R_{ijkl} : If x_{p1} is A_i and x_{p2} is A_j and x_{p3} is A_k and x_{p4} is A_l
then $(x_{p1}, x_{p2}, x_{p3}, x_{p4})$ belongstoClass C_{ijkl} with $CF = CF_{ijkl}$,
$i = 1, 2, \ldots, K; \quad j = 1, 2, \ldots, K;$
$k = 1, 2, \ldots, K; \quad l = 1, 2, \ldots, K,$

$$\tag{11.27}$$

where R_{ijkl} is the label of the linguistic rule, A_i, A_j, A_k, and A_l are linguistic labels, C_{ijkl} is the consequent class (*i.e.*, one of the three classes), and CF_{ijkl} is the grade of certainty. The consequent class C_{ijkl} and the certainty CF_{ijkl} can be determined from the given patterns in a similar manner as in the case of two-class classification problems in the two-dimensional pattern space $[0, 1]^2$ (see Ishibuchi et al. [20] to [23]).

By using the six linguistic labels, we generated $6^4 = 1296$ linguistic rules from the given patterns. Therefore our problem is to select a small number of linguistic rules from those rules.

In the same manner as the last subsection, we applied the genetic algorithm to the rule selection problem. We employed the same parameter specifications of the genetic algorithm as in the last subsection. After 1000 iterations of the genetic algorithm, we obtained four linguistic rules that correctly classified 146 patterns (97.3% of the given 150 patterns). The selected rules are shown in Figure 11.15 using the linguistic labels and also in Figure 11.16 using the membership functions. In the last column labeled "# of patterns" in these figures, we show the number of correctly classified training patterns by each linguistic rule. From Figures 11.15 and 11.16, we can see that each of the selected linguistic rules classifies many patterns.

In order to examine the average performance of the genetic algorithm, we iterated the same computer simulation ten times by generating different initial populations.

The average results of these computer simulations are as follows:

The average number of selected rules: 6.2
The average number of correctly classified patterns: 146.5 (97.7%)

No.	x_1	x_2	x_3	x_4	Class	CF	# of patterns
1	DC	DC	S	DC	1	1.00	50
2	DC	DC	M	DC	2	0.79	47
3	DC	DC	DC	ML	3	0.70	36
4	DC	M	DC	L	3	1.00	13

FIGURE 11.15
Selected linguistic rules represented by linguistic labels.

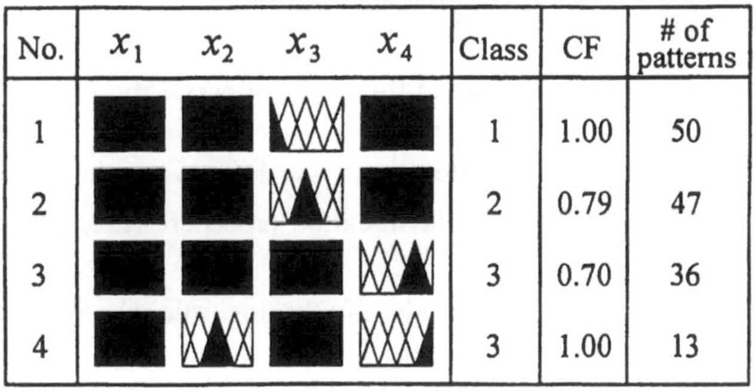

FIGURE 11.16
Selected linguistic rules represented by membership functions.

11.5 Extension to a Hybrid Genetic Algorithm

In the rule selection problem, we generated candidate linguistic rules by the rule generation procedure (*i.e.*, Procedure 1) in Section 11.2 where the grade of certainty of each rule was determined with no tuning procedure. In this section, first we briefly describe how the grade of certainty of each rule can be adjusted to improve the performance of a fuzzy-rule-based classification system. Then we propose a hybrid algorithm that incorporates a learning procedure of the grade of certainty into our genetic algorithm. It is shown by computer simulations on the iris data that a small number of linguistic rules with high classification power are selected by the proposed hybrid algorithm.

11.5.1 Adjustment of the Grade of Certainty

From the classification procedure (*i.e.*, Procedure 2) in Section 11.2, we can see that a pattern $x_p = (x_{p1}, x_{p2})$ is classified by the following linguistic rule $R_{\hat{i}\hat{j}}$:

$$\mu_{\hat{i}}(x_{p1}) \cdot \mu_{\hat{j}}(x_{p2}) \cdot CF_{\hat{i}\hat{j}} = \max_{R_{ij} \in S} \left\{ \mu_i(x_{p1}) \cdot \mu_j(x_{p2}) \cdot CF_{ij} \right\} \qquad (11.28)$$

If the consequent class $C_{\hat{i}\hat{j}}$ of this rule is the same as the true class of x_p, this pattern is correctly classified, otherwise x_p is misclassified. When x_p is misclassified by the linguistic rule $R_{\hat{i}\hat{j}}$, it is a natural strategy to decrease the grade of certainty of $R_{\hat{i}\hat{j}}$ as the punishment of the misclassification. On the contrary, when x_p is correctly classified by the linguistic rule $R_{\hat{i}\hat{j}}$, it is natural to increase the grade of certainty of $R_{\hat{i}\hat{j}}$ as the reward of the correct classification.

The above reward-and-punishment scheme can be written as the following learning algorithm (see Nozaki et al. [26]).

Procedure 3: learning procedure of the grade of certainty

Step 1: For each of the given patterns x_p, $p = 1, 2, \cdots, m$, do the following:

1. Classify x_p by the classification procedure (*i.e.*, Procedure 2) in Section 11.2.

2. Using Equation 11.28, identify the linguistic rule $R_{\hat{i}\hat{j}}$ that is responsible for the classification of x_p.

3. When x_p is correctly classified, increase the grade of certainty of the linguistic rule $R_{\hat{i}\hat{j}}$ in Equation 11.28 as

$$CF_{\hat{i}\hat{j}} := CF_{\hat{i}\hat{j}} + \eta_1 \cdot (1 - CF_{\hat{i}\hat{j}}), \qquad (11.29)$$

where η_1 is a positive learning constant.

4. When x_p is misclassified, decrease the grade of certainty of the linguistic rule $R_{\hat{i}\hat{j}}$ in Equation 11.28 as

$$CF_{\hat{i}\hat{j}} := CF_{\hat{i}\hat{j}} - \eta_2 \cdot CF_{\hat{i}\hat{j}}, \qquad (11.30)$$

where η_2 is a positive learning constant.

Step 2: If a prespecified stopping condition is not satisfied, return to Step 1. The number of iterations of this procedure is used as a stopping condition in this chapter.

In this procedure, the grade of certainty of each linguistic rule is always in the unit interval [0, 1] if the positive learning constants η_1 and η_2 are less than unity (*i.e.*, $\eta_1 < 1$ and $\eta_2 < 1$). Since usually there are much more correctly classified patterns than misclassified patterns, a larger value is assigned to η_2 than η_1. Therefore the learning constants should satisfy the inequality $0 < \eta_1 < \eta_2 < 1$. In computer simulations of this chapter, we specified η_1 and η_2 as $\eta_1 = 0.0001$ and $\eta_2 = 0.1$.

11.5.2 Hybrid Genetic Algorithm

The learning procedure of the grade of certainty (*i.e.*, Procedure 3) can be incorporated into our genetic algorithm. Since Procedure 3 can be applicable to any rule set S, we apply it to all rule sets generated by the genetic operations. Our hybrid genetic algorithm can be written as shown in Figure 11.17 where the genetic operations are the same as in Section 11.3.

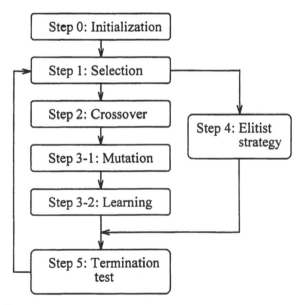

FIGURE 11.17
A hybrid genetic algorithm.

11.5.3 Simulation Results

Using the same parameter specifications as in Section 11.4, we applied the hybrid genetic algorithm to the iris data. We used ten presentations of all the given patterns as the stopping condition of the learning procedure of the grade of certainty. This means that Procedure 3 was iterated ten times for each of the generated strings in the hybrid genetic algorithm.

After 1000 generations, we obtained nine linguistic rules that can correctly classify all the 150 patterns (classification rate: 100%). The selected rules are shown in Figure 11.18.

No.	x_1	x_2	x_3	x_4	Class	CF	# of patterns
1	DC	DC	S	DC	1	1.00	50
2	DC	DC	M	M	2	0.95	40
3	DC	M	M	DC	2	0.79	8
4	M	MS	DC	DC	2	0.36	2
5	DC	MS	ML	ML	3	0.83	14
6	DC	M	DC	L	3	1.00	22
7	M	DC	ML	DC	3	0.53	9
8	M	MS	ML	DC	3	0.53	1
9	ML	DC	DC	DC	3	0.55	4

FIGURE 11.18
Selected linguistic rules.

In order to examine the average performance of the hybrid genetic algorithm, the same computer simulation with a different initial population was iterated ten times. We had the following average results:

The average number of selected rules: 10.7
The average number of correctly classified patterns: 149.8 (99.9%)

From the comparison of these results with those of the genetic algorithm in Section 11.4 (*i.e.*, 6.2 rules and 146.5 patterns on the average), linguistic rules selected by the hybrid genetic algorithm have higher classification power than those by the genetic algorithm.

11.6 Concluding Remarks

In this chapter, we proposed a genetic-algorithm-based method for selecting a small number of linguistic rules for classification problems. Then the proposed method was extended to a hybrid algorithm by incorporating a learning procedure into the genetic algorithm. By computer simulations on a numerical example, we demonstrated that the selected linguistic rules by the genetic algorithm coincided with our intuition very well. We also demonstrated that a small number of linguistic rules were selected by the genetic algorithm for the iris data. The selected linguistic rules correctly classified 146.5 patterns on the average (97.7% of the given 150 patterns). It was also demonstrated that linguistic rules selected by the hybrid algorithm have higher classification power than those by the genetic algorithm. The selected linguistic rules by the hybrid algorithm correctly classified 149.8 patterns on the average (99.9% of the given 150 patterns).

Since a small number of linguistic rules can be selected in our approach, we can carefully examine each of the selected rules. Careful rule examination is practically impossible if hundreds of rules are included in a fuzzy-rule-based classification system. In our approach, the classification knowledge is obtained in a form that human decision-makers can easily understand, that is, it is obtained as a compact set of linguistic rules. The clarity of the obtained rule set is an advantage of our approach as a knowledge acquisition tool.

References

[1] Sugeno, M., An introductory survey of fuzzy control, *Inf. Sci.*, 36, 59, 1985.

[2] Lee, C.C., Fuzzy logic in control systems: fuzzy logic controller. I and II, *IEEE Trans. Syst. Man Cybern.*, 20(2), 404, 1990.

[3] Takagi, T. and Sugeno, M., Fuzzy identification of systems and its applications to modeling and control, *IEEE Trans. Syst. Man Cybern.*, 15(1), 116, 1985.

[4] Wang, L.X. and Mendel, J.M., Generating fuzzy rules by learning from examples, *IEEE Trans. Syst. Man Cybern.*, 22(6), 1414, 1992.

[5] Sugeno, M. and Yasukawa, T., A fuzzy-logic-based approach to qualitative modeling, *IEEE Trans. Fuzzy Syst.*, 1(1), 7, 1993.

[6] Ichihashi, H. and Watanabe, T., Learning control by fuzzy models using a simplified fuzzy reasoning, *J. Jpn. Soc. Fuzzy Theory Syst.*, 2(3), 429, 1990 (in Japanese).

[7] Nomura, H., Hayashi, I., and Wakami, N., A learning method of fuzzy inference rules by descent method, *Proc. IEEE Int. Conf. Fuzzy Syst.* (San Diego, CA, March 8–12), p.203, 1992.

[8] Takagi, H. and Hayashi, I., NN-driven fuzzy reasoning, *Int. J. Approx. Reason.*, 5, 191, 1991.

[9] Lin, C.T. and Lee, C.S.G., Neural-network-based fuzzy logic control and decision system, *IEEE Trans. Comput.*, 40(12), 1320, 1991.

[10] Jang, J.S.R., Self-learning fuzzy controllers based on temporal back propagation, *IEEE Trans. Neural Networks*, 3(5), 714, 1992.

[11] Horikawa, S., Furuhashi, T., and Uchikawa, Y., On fuzzy modeling using fuzzy neural networks with the back-propagation algorithm, *IEEE Trans. Neural Networks*, 3(5), 801, 1992.

[12] Holland, J.H., *Adaptation in Natural and Artificial Systems*, University of Michigan Press, Ann Arbor, 1975.

[13] Goldberg, D.E., *Genetic Algorithms in Search, Optimization, and Machine Learning*, Addison-Wesley, Reading, MA, 1989.

[14] Karr, C.L., Design of an adaptive fuzzy logic controller using a genetic algorithm, in *Proc. 4th Int. Conf. on Genetic Algorithms*, San Diego, July 13–16, 1991, 450.

[15] Karr, C.L. and Gentry, E.J., Fuzzy control of pH using genetic algorithms, *IEEE Trans. Fuzzy Syst.*, 1(1), 46, 1993.

[16] Nomura, H., Hayashi, I., and Wakami, N., A self-tuning method of fuzzy reasoning by genetic algorithm, in *Proc. 1992 Int. Fuzzy Systems and Intelligent Control Conf.*, Louisville, March 16–18, 1992, 236.

[17] Fukuda, T., Ishigami, H., Shibata, T., and Arai, F., Structure optimization of fuzzy neural network by genetic algorithm, in *Proc. 5th IFSA Congr.*, Seoul, Korea, July 4–9, 1993, 964.

[18] Thrift, P., Fuzzy logic synthesis with genetic algorithms, in *Proc. 4th Int. Conf. on Genetic Algorithms*, San Diego, July 13–16, 1991, 509.

[19] Valenzuela-Rendon, M., The fuzzy classifier system: a classifier system for continuously varying variables, in *Proc. 4th Int. Conf. on Genetic Algorithms*, San Diego, July 13–16, 1991, 346.

[20] Ishibuchi, H., Nozaki, K., and Tanaka, H., Distributed representation of fuzzy rules and its application to pattern classification, *Fuzzy Sets Syst.*, 52, 21, 1992.

[21] Ishibuchi, H., Nozaki, K., and Tanaka, H., Efficient fuzzy partition of pattern space for classification problems, *Fuzzy Sets Syst.*, 59, 259, 1993.

[22] Ishibuchi, H., Nozaki, K., and Yamamoto, N., Selecting fuzzy rules by genetic algorithm for classification problems, in *Proc. 2nd IEEE Int. Conf. on Fuzzy Systems*, San Francisco, March 28–April 1, 1993, 1119.

[23] Ishibuchi, H., Nozaki, K., Yamamoto, N., and Tanaka, H., Construction of fuzzy classification systems with rectangular fuzzy rules using genetic algorithms, *Fuzzy Sets Syst.*, 65, 237, 1994.

[24] Fisher, R.A., The use of multiple measurements in taxonomic problems, *Ann. Eugenics*, 7, 179, 1936.

[25] Syswerda, G., Uniform crossover in genetic algorithms, in *Proc. 3rd Int. Conf. on Genetic Algorithms* (George Mason University, June 4–7, 1989), Morgan Kaufmann, San Mateo, CA, 1989, 2.

[26] Nozaki, K., Ishibuchi, H., and Tanaka, H., Trainable fuzzy classification systems based on fuzzy if-then rules, in *Proc. 3rd IEEE Int. Conf. on Fuzzy Systems*, Orlando, FL, June 26–29, 1994, 498.

12

A Genetic Algorithm Method for Optimizing the Fuzzy Component of a Fuzzy Decision Tree

Cezary Z. Janikow

Abstract   Fuzzy decision trees exploit popularity of decision tree algorithms for practical knowledge acquisition and representative power of fuzzy representation. They are extensions of symbolic trees, with tree-building routines modified to utilize fuzzy instead of strict domains, and with new inferences combining fuzzy interpolation and defuzzification with inductive methodology. Such fuzzy trees have been recently proposed and described. In this chapter, we describe a method for optimizing the fuzzy component of knowledge represented in the fuzzy tree. That is, we optimize the knowledge by adjusting the fuzzy sets for both linguistic terms and for decisions, and by selecting an optimal norm for combining conjunctions of fuzzy restrictions. A genetic algorithm is used to perform the highly constrained optimization. A simple example is used to illustrate the technique.

12.1   Introduction

Decision-tree algorithms provide one of the most popular methodologies for symbolic knowledge acquisition. The resulting knowledge, in form of symbolic decision trees and easily understood inference mechanisms, has been praised for its compressibility. This appeals to a wide range of users who are interested in domain understanding, classification capabilities, or the symbolic rules that may be extracted from the tree [38] and subsequently used in a rule-based decision system. This interest, in turn, has generated extensive research efforts resulting in a number of methodological and empirical advancements [24, 36, 37, 38]. The decision tree—or decision-tree classifier—approach was popularized by Quinlan [35] with the ID3 program. Systems based on this approach work well in symbolic domains. By principle, decision trees assign symbolic decisions to events. This makes them inapplicable in cases where a numerical decision is needed, or when the numerical decision improves subsequent processing [7]. As a result, trees suffer from decision instability. This problem is amplified when numerical domains are discretized to symbolic domains, since a small measurement error can now ignite instability.

Fuzzy sets are one of the most popular methods designed to overcome such limitations of symbolic systems. They provide bases for fuzzy representation. The ideas have been most readily translated into rule bases, resulting in fuzzy rules. Such fuzzy rule-based representation allows modeling language-related uncertainties, while providing a symbolic framework for knowledge comprehensibility. In this setting, the symbolic rules provide for ease of understanding and/or transfer of high-level knowledge, while the gradual fuzzy definitions, along with the available norms and composition mechanisms, provide the ability to model fine knowledge details [3, 19, 24]. Accordingly, fuzzy representation is becoming increasingly popular in dealing with problems of uncertainty, noise, and inexact data [24]. It has been successfully applied to problems in many industrial areas [19]. Most research in applying this new representative framework to existing methodologies has concentrated only on emerging areas such as neural networks and genetic algorithms [19]. For example, these two are being used to optimize fuzzy definitions an to learn fuzzy rules [8, 12, 21, 24, 39, 42]. However, little has been done to combine fuzzy ideas with symbolic AI, or the popular decision trees, in particular. Fuzzy decision trees were designed to combine these two methodologies in such a way that the comprehensibility be preserved, but the representative power be increased to model language uncertainties into gradual response curves [13, 14, 15].

Decision trees are made of two major components: a procedure to build the symbolic tree and an inference for decision making. Additional procedures can be utilized to restructure the tree in order to improve its generalization properties. At the moment, the first two of the three procedures have been translated into the new representation. In addition, some of the most important methodological advancements available in symbolic decision trees have been incorporated. We are currently working on additional inferences for fuzzy trees and on knowledge optimization methods. This chapter presents our current results in optimizing the fuzzy knowledge component. This is done in two stages: statically before the tree is built and dynamically in the tree-building routine. Symbolic component optimization, that is, tree restructuring [31], is left for future research.

For a complete presentation we first brief symbolic decision trees in Section 12.2 and fuzzy representation in Section 12.3. In Section 12.4 we introduce genetic algorithms, which are used in the optimization. In Section 12.5 we overview the tree-building routine for fuzzy decision trees, and subsequently modify it to incorporate the optimization. In Section 12.6 we present details of the GA used for this highly constrained optimization. Our method, in addition to being incorporated into the tree-building routine, differs from other previous GA optimizations of fuzzy sets [20] by utilizing the recently proposed method for constraints optimization [26, 27]. Finally, in Section 12.8 we present a small illustrative experiment.

12.2 Decision Trees

In *supervised learning*, an example, also called an *event* is represented by a conjunction of feature descriptions and decision assignment. The objective is to induce decision procedures with *discriminative* (most cases), *descriptive*, or *taxonomic* bias [30]. Following the *comprehensibility principle*, which calls upon the decision procedures to use language and mechanisms suitable for human interpretation and understanding [30], symbolic systems remain extremely important in many applications and environments. Among such, decision trees are one of the most popular.

In decision trees, such as the ID3 program and its derivative C4.5 [38], decisions are symbolic. Attributes must also have symbolic features. If an attribute is not symbolic to start with, it must be discretized. This discretization can be dynamic while building the tree, by computing discriminative (usually binary) thresholds [2]. Alternatively, the discretization can be conducted statically, that is, independently on all numeric attributes prior to tree construction. This is often done to increase comprehensibility of the induced language, when some background or common sense knowledge proposes meaningful discretization scheme, or when the language is fixed in an existing knowledge system. In this work, we assume this latter case since our objective is to use comprehensible features. Dynamic discretization, or optimization of the available language, is planned at the end of this project—hopefully to lead into future projects.

The objective of a symbolic decision-learning system is to provide an inference for classifications of all events—that is, to generalize the partial knowledge represented in the initial data base of events. In other words, the procedure is to partition the event space with different decisions. Ideally, this should be an equivalence partition with large blocks—they increase comprehensibility of the generated knowledge and usually improve generalization properties. However, in general, the generated partition may not be *complete* nor *consistent*. This may be due to insufficient language, errors or noise in data, impression in measurements, or other uncertainties.

Decision-tree classifier systems are made up of two different procedures: one to build the tree, another for knowledge inference. Some may, in addition, reconstruct the generated knowledge, by possibly pruning the tree to improve its generalization properties [31]. Others may convert the tree to a set of rules, and subsequently use a rule-based inference method [38]. Each *training* instance contains a set of features and the associated classification. Each internal node is labeled with an attribute and has outgoing edges corresponding to domain values of the attribute. Each node in the tree contains a subset of the training events—those that satisfy the conditions leading to the node. Figure 12.1 illustrates a typical decision tree. In this figure, two decisions are illustrated as white and black. Internal nodes, without unique decisions, are gray. The leaf node L_2 contains training events with the black decision, which also contain the features: *Color* is *blue* and no *Textured*. Implication of this structure and the inference is that any case with these two particular features will be assigned the same black decision.

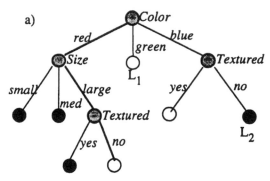

FIGURE 12.1
An example of a decision tree.

The tree-building is a recursive partitioning routine. The root of the decision tree contains all training events, and it is recursive split with the events partitioned. At each node, the splitting stops when its events represent a unique decision, all attributes are used up on the path or when some other user-specified criteria are met. Then the node becomes a leaf and it assumes classification of its events. When it is decided to further split the node, one of the attributes is selected, with its symbolic features splitting the node and partitioning its events. One of the most popular attribute selection mechanisms is one that maximizes information gain [32]. This mechanism is computationally simple as it assumes independence of attributes. However, empirical evaluations indicate the quality of this trade-off. This procedure is based on information contents and can be summarized as follows:

1. Compute the information content at node P, given by

$$I_P = -\sum_{c_i}^{|C|} p_i \times \log p_i$$

where $|C|$ is the number of decisions and p_i is the probability of decision c_i in the node's events.

2. For each remaining attribute a_i and each of its symbolic domain value a_{ij}, compute the information content $I_{p,a_{ij}}$, based on events that fall into the corresponding subnodes.

3. Expand node P using the attribute a_i that maximizes the information gain

$$I_P - \sum_j^{|D_i|} w_j \times I_{p,a_{ij}}$$

where w_j is the relative weight of events at node P having the a_{ij} feature and D_i is the symbolic domain of the attribute.

The inference for an event e assigns the decision of the training events from the leaf node whose feature restrictions on the path are satisfied by the event. Problems arise when such a node does not exist, there is a number of such nodes, or the decision at the leaf is not unique. These problems have been addressed in the context of decision trees, and a number of potential solutions have been proposed [2, 5, 24, 31, 32, 36, 37, 38].

One major source of such problems are continuous attributes and the noise in their uncertainty/noise measurements. To deal with continued values, in general, dynamic discretization has been used to threshold the numerical domains into *buckets*. Dynamic refers to dividing the domain separately for different nodes of the tree, while it is being constructed. The most often used is a single split, resulting in binary trees. A drawback of this approach is that a small noise in the measurement, or even inaccuracy stemming from the training data, can easily fool the system into erroneous decisions [24, 36]. A way to reduce the problem is to allow the buckets to overlap. This, however, aggregates another problem—an event presented for classification is now likely to satisfy multiple leaves. Moreover, to differentiate between a full match and a match based on the overlap, the match may be defined as partial [36]. Similarly, missing values can force multiple partial matches [37]. In addition, nondiscriminative language, or tree pruning, may lead to leaves with nonunique decisions. One way to treat such is to view them as probabilistic classifiers, and to combine such probabilities from multiple leaves [36].

Another drawback of this dynamic discretization is a reduced comprehensibility—the selected splits might be difficult to comprehend (e.g., *Fever* < 102.39), and a restriction based on a given attribute may appear a number of times for a single leaf node (leading to larger trees). When this becomes a burden, a static discretization scheme can be used, which creates the divisions prior to building the tree. This method has been often used in machine learning applications.

Fuzzy sets, with subsequently used fuzzy inferences, can be seen as a direct extension of such previous attempts to introduce and process the overlapping buckets, but with a more rigorous method, which draws on the extensive research of fuzzy rules and fuzzy control. In our case, we define the linguistic domains *a priori*, following the comprehensibility principle, with some initial fuzzy sets. These are subsequently optimized with a genetic algorithm. This optimization is both static and dynamic, with the latter being more restrictive to overcome the mentioned drawbacks.

12.3   Fuzzy Sets, Rules, and Approximate Reasoning

In classical set theory, an element either belongs to a certain set or does not. In this case, Boolean logic can be employed resulting in a *crisp* system. However, in the real world, such a scenario is often unrealistic because of insufficient language, imprecise measurements, and so on. It is widely believed that some numerical components must be employed to accommodate such problems [5]. However, purely quantitative approaches lack other characteristics desired of a successful approach. In particular, the most often-cited disadvantage is lack of human-level comprehensibility and hence difficulties with transfer and other usage of knowledge [30]. Therefore, a successful learning system must combine quantitative and qualitative representation and reasoning. Accordingly, neural networks are being pushed toward the symbolic level by exploration of topological structures and design of new paradigms [39], decision trees are being designed to accommodate probabilistic measures [36], and decision rules are being extended by flexible matching components or other methods for uncertainty propagation [29]. At the same time, nonbinary extensions to sets are being proposed.

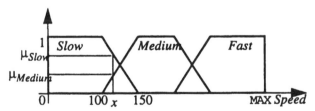

FIGURE 12.2
Fuzzy subsets for *Speed*, and the memberships for some actual speed.

In fuzzy sets, a fuzzy subset A of the universe of discourse U is described by a membership function $\mu_A(x) : x \in U \rightarrow [0, 1]$, which represents the degree to which x belongs to the set A. A fuzzy linguistic variable is an attribute whose domain contains linguistic terms—labels for fuzzy subsets. For example, consider a continuous variable *Speed*. This attribute becomes fuzzy when its numerical domain is replaced with fuzzy linguistic labels *Slow, Medium, Fast*, each of which represents a fuzzy subset of the universe of possible speed values. In Figure 12.2, the actual speed x belongs to both the *Slow* and the *Medium* subsets with different degrees. Using trapezoidal definitions, a fuzzy linguistic value can be represented by a vector of the four corners of the trapezoid. For example, the *Slow* value in Figure 12.2 could be represented as [0, 0, 100, 150]. Trapezoidal, or triangular, definitions tend to be the most popular due to computational efficiency and empirical quality, but other *smoother* forms are being investigated as well [41] (they produce smoother responses).

Fuzzy sets form the basis for extending symbolic *if–then* rules to fuzzy rules of the form [24]:

$$R^k : \text{ if } P_1 \wedge P_2 \ \ldots \ \text{ then } C^k$$

where P_i is a restriction on a fuzzy variable V_i (*e.g.*, *Speed* is *Medium*), and C^k is the fuzzy consequent—a restriction on the decision variable V_C. While, in general, the variables of the antecedent and of the consequent may come from the same set (*e.g.*, in planning systems with rule chaining), for many interesting problems (*e.g.*, decision or control problems) these come from disjoint sets, often with a unique decision variable V_C. We will continue our discussion under this assumption since our aim is to develop a decision-making mechanism. A fuzzy rule-based system is a set of such fuzzy rules $\{R^k\}$, along with special inference procedures.

To determine the fuzzy decision of such a system, current data (facts, measurements, or other observations) are used in an analogous way to those of other rule-based systems. That is, the strength of each rule must be determined and a conflict resolution used to compute the final decision. To determine these, four evaluation mechanisms must be defined for the fuzzy rule-based system:

1. f_0 to determine how a single data value x_i for variable V_i satisfies a fuzzy restriction $P_i : [V_i \text{ is } v_i]$,

2. f_1 to combine levels of satisfactions of fuzzy restrictions of the conjunctive antecedent,

3. f_2 to propagate the activation of the antecedent to the consequent,

4. f_3 for the conflict resolution from multiple consequents.

In fuzzy rules, f_0 is identical to the previously defined $\mu_{v_i}(x) \cdot f_1$, which evaluates a conjunction of fuzzy restrictions, most often identical to the minimal or the product operator. f_2 is usually similarly defined. Finally, f_3, which evaluates a disjunction of different decisions, is usually defined with the maximal or the sum operator [19, 33]. The center-of-gravity method has been developed for defuzzification to crisp responses. It simply states that the quantitative response is the center of gravity of the resulting fuzzy set. If the maximal operator is used for f_3, this method would simply produce the center of the union of all consequents. However, if the sum is used for f_3, the method produces the center of gravity of all accumulating evidence (this latter computation method is often implicitly assumed in the name). In this case, the response is computed as

$$\delta = \sum w_k \theta_k \alpha_k \zeta_k / \sum w_k \theta_k \alpha_k$$

where θ_k represents the activation degree (result of f_2) of consequent C^k, α_k and ζ_k are the area and centroid of the linguistic consequence, δ is the resulting decision, and w are possible weights to indicate rule strengths. The summation takes place over all rules in the base. If it is necessary to determine the linguistic response, as it would be to test the system using test data with linguistic decisions, it is possible now to apply some operators to reverse the centroid mechanism. One simple solution is to select $max_{c \in V_c} [\mu_c(\delta)]$.

12.4   Genetic Algorithms

In recent years, problem solving has begun to emerge as interactions of active agents with the environment and the surrounding world rather than by isolated operations. Some of these ideas are derived from nature, where organisms both cooperate and compete for resources of the environment in the quest for a better adaptation. Such observations led to the design of algorithms which simulate these natural processes. The genetic algorithm (GA) represents one of the most successful approaches. GAs are adaptive search methods that simulate some of the natural processes: selection, information inheritance, random mutation, and population dynamics. The principles were first elucidated in Reference [11], and since then the field has matured and enjoyed many successful applications [6, 10]. At first, GAs were most applicable to numerical parameter optimizations due to an easy mapping from the problem to representation space. Today, they find more and more general applications due to better understanding of the necessary properties of the required mapping and new methodologies to process problem constraints.

A GA operates as a simulation in which individual agents, organized in a population, compete for survival and cooperate to achieve a better adaptation. The agents are called *chromosomes*. The chromosome structure (*genotype*) is made up of genes. The meaning of a particular chromosome (*phenotype*) is defined externally by the user in such a way that a complete chromosome represents a potential solution to a problem at hand. Traditional genetic algorithms operate on strings of bits.

GAs use two mechanisms to provide for the adaptive behavior: selective pressure and information inheritance. Selection, or competition, is a stochastic process with survival chances of an agent proportional to its adaptation level. The adaptation is measured by evaluating the phenotype in the problem environment. This selection imposes a pressure promoting survival of better individuals, which subsequently produce offspring. Cooperation is achieved by merging information usually from two agents, with the hope of producing more adapted individuals (better solutions). This is accomplished by *crossover*. The merged information is inherited by the offspring. Additional *mutation* aims at introducing extra variability. Algorithms utilizing these mechanisms exhibit great robustness due to their ability to maintain an adaptive balance between efficiency and efficacy. The simulation is achieved by iterating the basic steps of evaluation, selection, and reproduction, after some initial population in generated. The initial population is usually generated randomly, but some knowledge of the desired solution may be used to an advantage. The iterations continue until some resources are exhausted. For example, the simulation may be set for a specific time limit or a fixed number of iterations. Alternatively, if some information about the sought solution is available, the simulation may continue until some criteria are met. Finally, the population dynamics may be observed and the simulation may stop if convergence to a solution is detected.

Evaluation is performed by a task-specific evaluation function. Stochastic selection (with replacement) is applied to the beginning population instance, producing the intermediate state. Because of the selective pressure favoring survival of better fitted individuals, the average fitness of the chromosomes increases. However, no new individuals appear. Following the selection, reproduction operators are applied to members of the intermediate population. In this process, some chromosomes are modified. Therefore, the new population instance will finally contain some new chromosomes. This process continues for a number of iterations. The described iterative model is called the *generational* GA. Variations of this model are often used instead [6].

If generic crossover and mutation operators are used, the only relation on the process at hand is the evaluation function providing the simulated environment. This is a great advantage, leading to domain-independent characteristics of the algorithm. This is also a great limitation, prohibiting utilization of any available information about the problem.

In many applications, an additional difficulty is introduced by constraints that the sought solution must satisfy. Constraints cause most serious problems in traditional models where the search space is spanned by a fixed representation (*i.e.*, binary), while the feasible space is only a portion of the former. This may cause the search to drift into improper regions of the search space. There are a number of approaches to deal with the problem. The simplest is to throw away any infeasible offspring [23], but this may quickly become ineffective when the constraints become more severe. In the 1980s, the most often proposed solution was to modify the evaluation to penalize infeasible solutions [39]. This can often work nicely for so-called *weak* constraints, that is, constraints that do not necessary invalidate solutions when violated, but generally prove disastrous for *strong* constraints, which are often present in function optimization. Another approach is to provide special repair algorithms that will move any inconsistent offspring into the feasible space. Again, this may prove too inefficient in highly constrained problems.

The two most appealing approaches are to modify the representation and/or to modify the operators. If it is possible to devise a different representation that spans only the feasible space, any offspring is guaranteed to satisfy the constraints. Two obvious difficulties are to find such representation, which is highly problem and constraint specific, and to possibly provide new operators to work in this representation. A good example of this approach is given in Reference [16]. Fortunately, in some cases it might be possible to use the existing generic representation and provide operators that are closed in the feasible subspace of the spanned representation space. This idea was given in References [26] and [27] and became often used for problems of parameter optimization with strong linear constraints.

12.5   Fuzzy Decision Trees

12.5.1   Fuzzy Event Representation

We use the quantitative representation of attribute values. Here, we assume that all attributes have numerical domains. If an attribute is symbolic *linear* (*i.e.*, has similarity measures forming total ordering on the symbolic values, as in *Size* = { *Small, Medium, Large* }), the values will represent centroid values of appropriate fuzzy definitions. If an attribute is *nominal*, the definitions are disjoint. The same can be applied to decisions. This quantitative-input approach is actually advantageous since it allows for the natural handling of uncertainty and noise in determination of these symbolic values. As an example, consider the event space spanned by symbolic attributes *EducationLevel* = { *Elem, HS, BS, Grad* } and *Employed* = { *Yes, No* }. Since it is often difficult to assign one of these features to a person (for example, some college and a sporadic employment), it would instead be advantageous to use the fuzzy definitions illustrated in Figure 12.3. The same can be applied to decisions: if there is any question about the decision, the definitions can be overlapping. Due to interpolative nature of the center-of-gravity inference, we assume that multivalued decisions have the linear characteristics.

We also allow the events to be weighted with respect to our confidence (all weights may be assumed 1 if unused). For example, assuming the two attributes of Figure 12.3 and decision *Credit-Worthy*, a part-time (about 20 hr) employee who attended some graduate school may be given high credibility. This event may be represented as $[EducationLevel = 0.8][Employment = 0.5] \Rightarrow [CreditWorthy = 0.9]$; weight = 1.00. Note that, if necessary, the events may be described using some of the linguistic labels.

12.5.2   Assumptions and Notation

The decision tree will be built at some comprehensible symbolic level. Therefore, symbolic quantization of the numerical domains must be defined. For illus-

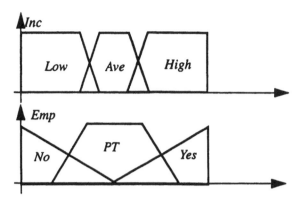

FIGURE 12.3
Sample fuzzy definitions.

trations, we use the static fuzzy trapezoidal definitions. A trapezoidal definition can be expressed as a quadruple of its horizontal corner coordinates since they all share the same height: $[x_1, x_2, x_3, x_4]$, where $x_1 \leq x_2 \leq x_3 \leq x_4$. Moreover, due to computational needs to process areas, we require that $x_1 < x_4$. However, all the subsequent discussion and processing is transparent of the kind of fuzzy definitions used, except for the actual implementation of the optimization routine.

We assume that all features are given in normalized to [0, 1] quantitative forms. Symbolic features are replaced by numeric values fitting dummy definitions. For example, for the inherently symbolic nominal attribute *isMale*, the dummy definitions might be $[-0.1, 0, 0, 0.1]$ and $[0.9, 1, 1, 1.1]$ (the interval is introduced to ensure the definition areas are nonzero). Then, each *False* value would be represented by the number 0.0 and each *True* value would be represented by the number 1.0. Decisions are treated as ordinary attributes. That is, they will have dummy definitions if such are not given, and they will be represented by numerical values. Any numerical value outside of the normalized domain indicates an unknown value. We do not allow unknown decisions in the training set.

For now, the tree is not pruned. Because of that, each internal node has a branch for each value of its attribute's label, except when no events satisfy the restriction. To provide for future pruning extensions, we assume that during the inference, an unknown value in a node occurs either when the attribute-value of the event to be classified contains an unknown value or the node has no branch for it. Before we define the procedures for the fuzzy-build and the fuzzy-inference, let us introduce subsequently used notation.

1. The set of attributes is denoted by $A = \{a_1, \ldots, a_n\}$.

2. For each attribute $a_i \in A$

    * D_i denotes the linguistic domain set; it contains labels representing fuzzy definitions, *e.g.*, $D_{Size} = \{Small, Med, Large\}$.

    * d_i denotes the normalized numerical $[0, 1]$.

    * v_p^i denotes the pth linguistic value of attribute a_i; it is the label for the pth fuzzy set; *e.g.*, $v_1^{Size} = Small$.

    * v_p^{-i} denotes the restriction: a_i is v_p^i.

3. Weights are denoted by W; their normalized domain is denoted by d_W.

4. The set of fuzzy decisions is denoted by D_c; d_c is the normalized numerical domain.

5. For each node N of the fuzzy decision tree

    * F^N denotes the set of restrictions on its path (with conjunctive interpretation), *e.g.*,

        $$F^{L2} = \{v_{blue}^{-Color}, v_{no}^{-Textured}\} \text{ in Figure 12.1.}$$

    * A^N is the set of used up attributes on its path, *i.e.*, it contains attributes whose restrictions appear in F_N.

    * $N \diamond a_i$ denotes the set of its children nodes using the attribute $a_i \in (A - A^N)$ for split.

    * $N \bullet v_p^i$ denotes the particular child node of N using attribute a_i for split and linguistic value $v_p^i \in D_i$ for the branch

        $$F^{N \bullet v_p^i} = F^N \cup v_p^{-i}$$

    * P_k^N denotes its event count for decision $v_k^c \in D_C$.

    * $P_k^{N \bullet v_p^i}$ denotes its event count for decision $v_k^c \in D_C$ which also satisfies the restriction v_p^{-i}.

        $$P_k^N \le \sum_p^{|D_i|} P_k^{N \bullet v_p^i}$$

        due to more relaxed fuzzy matching.

    * P^N denotes its total event count; I^N denotes its information count.

- $I^{N \Diamond a_i}$ denotes weighted information count in its children after using a_i for split, and $G_i^N = I^N - I^{N \Diamond a_i}$ is the associated gain on attribute a_i at the node N. Note that to maximize gain the weighted information count must be minimized.

6. The set of events is E, with each $e_j \in E$ described by numerical features, a numerical decision assignment, and a numerical weight.

   - $e_j = (x^l, \ldots, x^n, y_j, w_j) \in E$, where $(x^l \times \ldots x^n \times y \times w) \in (d_l \times \ldots d_n \times d_c \times d_w)$.

7. α denotes the area, ζ denotes the centroid of a linguistic value.

8. To deal with missing attribute values (we assume no missing decisions nor weights)

   - $x^i \notin d_i$ denotes a missing value for a_i.
   - $P^{N|(a_i \in d_i)}$ denotes weighted percentage of events in E^N with known values for attribute a_i.

$$P^{N|(a_i \in d_i)} = \frac{P^N - P^{N|(x_i \notin d_i)}}{P^N}$$

   - $k_j^{i,p}$ denotes the grade that event e_j satisfies the restriction v_p^i

$$k_j^{i,p} = \begin{pmatrix} \mu_{v_p^i}(x_j^i) & \text{if } x_j^i \in d_i \\ \left|\frac{1}{D_i}\right| & \text{otherwise} \end{pmatrix}$$

12.5.3 Procedure to Build a Fuzzy Decision Tree

The fuzzy decision tree is built using the symbolic linguistic labels for attribute and decision definitions. The traditional approach uses measures based on crisp numbers of elements in the partitioned sets. However, in our case, the elements are fuzzy and, therefore, ideas of fuzzy sets must be employed. Except for that, we follow the best strategies available for symbolic decision trees.

To modify the way in which the number of events belonging to a node is calculated, we adapt rules of inference from fuzzy logic, extended for our representation. Therefore, in addition to the methodology, we have to provide f_0 and f_1 functions. f_0 can trivially use μ. However, borrowing the existing methodology to deal with missing features in the training set, we will use $f_0(x_j^i, v_p^{-i}) = k_j^{i,p}$ instead. Then, consider the case of visiting node N during tree expansion. An event can be found in the node based on its match to the set of restrictions leading to the node. Since the interpretation of the set is conjunctive, we may use any standard method for conjunctive evaluations of antecedents (f_1). We proposed [15] a few options to evaluate these restrictions, following the most studied norms in fuzzy control [33]. Moreover, we proposed incremental versions of those for computational efficiency [18].

1. Compute event counts in node N, given the events it contains (for the *Root*, $W^{Root} = W$):

$$P_k^N = \sum_{j=1}^{|E|} w_j^N \cdot \mu_{v_k^c}(y_j)$$

$$P^N = \sum_{k=1}^{|D_c|} P_x^N = \sum_{k=1}^{|D_c|} \sum_{j=1}^{|E|} w_j^N \cdot \mu_{v_k^c}(y_j)$$

where w_j^N is the weighted set of events in node N, computed as follows:

- $w_j^{Root} = w_j$

- $w_j^{v_p^i} = w_j \cdot k_j^{i,p}$

- $w_j^{N \bullet v_p^i} = f_1(w_j^N, k_j^{i,p})$

2. Having these counts, we follow the standard information contents formula:

$$I^N = -\sum_{k=1}^{|D_c|} \frac{P_k^N}{P^N} \cdot \log \frac{P_k^N}{P^N}.$$

3. At each node, we must search the set of remaining attributes for the one that gives the maximal information gain, which is subsequently used to split the node. The process stops when:

    - $I^N \leq$ some threshold, or $P^N \leq$ some threshold, or $A^N = A$.

    If split is decided, the split attribute is selected from all remaining attributes $(a_i \in A - A^N)$ as follows:

    - Reject a_i from consideration based on chi-square test of independence [24].

    - For each $v_p^i \in D_i$, calculate $I^{N \bullet v_p^i}$.

    - Calculate $I^{N \Diamond a_i}$, the formula for information based on attribute a_i, which is reduced by weighted percentage of events in N that have unknown values for a_i [37].

    - Compute G_i^N, the information gain for attribute a_i, which may also be reduced for excessive domain sizes [35].

    - Select attribute a_i such that reduced information gain G_i^N is maximal.

- Split the node into $|D_i|$ subnodes, with the set of events in each subnode defined by weights of step. The only exception is that subnodes that do not contain any events are folded.

12.5.4 Static Optimization

Static optimization refers to optimization prior to tree construction. As explained in Section 12.2, this method increases comprehensibility of the generated tree. Because in this case there are no rules yet, previous GAs or other methods cannot be applied [20, 42]. Instead, we introduce an auxiliary variable with restrictions induced by conjunctions of existing restrictions. Therefore, the corresponding fuzzy sets will be dependent on fuzzy sets for the original attributes and on the f_1 norm (again, because we do not view these subspaces as rules, we do not optimize f_2). Using this variable, we measure the information contents of the data after using this variable for a split. This quantity is then minimized with a GA, with the original fuzzy sets and the choice for f_1 being the dependent variables. This minimization is highly constrained due to specific requirements imposed on the trapezoidal (or any other) fuzzy sets. The objective of this static optimization is to increase discrimination at the smallest possible discernibility level spanned by the language of A. It is important to realize that such optimization does not guarantee a reduction in the size of the tree, even though it is likely. Instead, it is aimed at increasing consistency of the fuzzy tree (which implicitly should reduce its size since one of the stopping criteria for the tree-building routine is the current information contents).

Let us denote $\bar{a}$ to be a new induced attribute *s.t.*

$$d_{\bar{a}} = d_1 \times d_2 \times \ldots d_n = [0, 1]^n$$

$$D_{\bar{a}} = D_1 \times D_2 \times \ldots D_n$$

where $n = |A|$. Now we use this induced attribute to split the set of events E, with weights W. Following Section 12.5.3, the information contents after this split is defined by:

$$I^{\bar{a}} = \frac{\sum_{r=1}^{|D_{\bar{a}}|} P_r^{v_r^{\bar{a}}} \cdot I_r^{v_r^{\bar{a}}}}{\sum_{r=1}^{|D_{\bar{a}}|} P_r^{v_r^{\bar{a}}}}$$

where

$$P_k^{v_r^{\bar{a}}} = \sum_{j=1}^{|E|} w_j^{v_r^{\bar{a}}} \cdot \mu_{v_k^c}(y_j)$$

$$P^{v_r^{\bar{a}}} = \sum_{k=1}^{|D_c|} P_k^{v_r^{\bar{a}}} = \sum_{k=1}^{|D_c|} \sum_{j=1}^{|E|} w_j^{v_r^{\bar{a}}} \cdot \mu_{v_k^c}(y_j)$$

and

$$I^{v_{\hat{a}}_r} = -\sum_{k=1}^{|D_c|} \frac{P_k^{v_{\hat{a}}_r}}{P^{v_{\hat{a}}_r}} \cdot \log \frac{P_k^{v_{\hat{a}}_r}}{P^{v_{\hat{a}}_r}}.$$

Suppose we look at the elementary block $r = v_{p_1}^1 \times v_{p_2}^2 \ldots v_{p_n}^n$ induced by the auxiliary attribute.

Then, $I^{\bar{a}}$ is a function of $\left(w_j^{v_{\hat{a}}_r}, \mu_{v_k^c}(y_j) \right)$ for all $e_j \in E$. However, $w_j^{v_{\hat{a}}_r} = w_j \times f_1 \left(k_j^{1,p_1}, k_j^{2,p_2}, \ldots, k_j^{n,p_n} \right)$. Therefore, the information contents of this auxiliary variable is a function of weights W and all of the fuzzy sets defining terms for attributes of A. Moreover, it is also a function of the fuzzy sets for the decisions and of the composition norm f_1. W is a constant external quantity (in other words, we do not wish to only look at those samples that look nice—we need to consider all of the available samples). Therefore, minimization of the information contents derived by $I^{\bar{a}}$ can be carried out by (1) adjustments of fuzzy sets for terms defining attributes, (2) adjustment of fuzzy sets for decision terms, and (3) selection of f_1. To preserve some comprehensibility, we assume that the number of fuzzy terms, that is, sizes of all domains D, is fixed. If desired, this restriction can be relaxed. These adjustments and selections can be carried out simultaneously if a robust optimization mechanism is available. We use genetic algorithms. The only change that is required in Section 12.5.3 for the tree-building procedure is to start with an additional task:

> 0. Optimize the initial fuzzy sets and the norm f_1 using the GA, by minimizing $I^{\bar{a}}$.

12.5.5 Dynamic Optimization

The principles of Section 12.5.4 can be partially extended for dynamic optimization, that is, optimization while constructing the tree. Of course, at this moment we cannot modify the fuzzy sets, defining terms for decisions nor for attributes previously used in the tree. Therefore, the complexity of this optimization decreases while descending the tree. In other words, this optimization is different from that of Section 12.5.4 by only the additional constraints

1. The fuzzy sets for decisions in D_C are constant.

2. The fuzzy sets describing terms for attributes in $U_{N \in \text{Leaves}} A^N$ are constant.

3. f_1 norm is constant.

This optimization is used dynamically while building the fuzzy tree by an additional modification of the routine of Section 12.5.3. This idea is based on the observation that the information contents of a node N, I^N, is a function of W (which is constant), fuzzy sets for decisions (which we assume are now constant for all paths in the tree), and fuzzy sets for attributes in A^N (which we also assume are now constant). Therefore, we may modify the procedure of Section 12.5.3 by preceding step 3 with another step

> 3a. Optimize the fuzzy sets of the remaining attributes in $A = U_{N \in \text{Leaves}} A^N$ by applying the GA to minimize the additionally constrained I^a.

12.6 GA Optimization Algorithm

This optimization requires to minimize $I^{\overline{v_r^a}}$. The optimization parameters are trapezoidal corners for all fuzzy sets (or other parameters if other shapes are used) plus the f_1 norm (if desired to select the optimal). This optimization is highly constrained. Moreover, additional parameters must be set constant while building the tree. Fortunately, all these constraints are linear. Here we present the methodology (GENOCOP) along with some constraints useful in this case.

12.6.1 Linear Constraints

An optimization problem with linear constraints can be described as – optimize a function $f(x_1, x_2, \ldots, x_q)$, subject to the following sets of linear constraints:

1. Domain constraints: $l_i \le x_i \le u_i$ for $i = 1, 2, \ldots, q$. We write $\vec{l} \le \vec{x} \le \vec{u}$, where $\vec{l} = \langle l_1, \ldots, l_q \rangle$, $\vec{u} = \langle u_1, \ldots, u_q \rangle$, $\vec{x} = \langle x_1, \ldots, x_q \rangle$.

2. Equalities: $A\vec{x} = \vec{b}$.

3. Inequalities: $C\vec{x} \le \vec{d}$.

12.6.2 Processing Ideas

Before we discuss processing details for our constraints, let us present a small example which should provide some insight into the proposed methodology. Let us assume we wish to minimize a function of six variables:

$$f(x_1, x_2, x_3, x_4, x_5, x_6)$$

subject to the following constraints:

$$x_1 + x_2 + x_3 = 5$$
$$x_4 + x_5 + x_6 = 10$$
$$x_1 + x_4 = 3$$
$$x_2 + x_5 = 4$$
$$x_1 \geq 0, x_2 \geq 0, x_3 \geq 0, x_4 \geq 0, x_5 \geq 0, x_6 \geq 0.$$

We can take an advantage from the presence of four independent equations and express four variables as functions of the remaining two:

$$x_3 = 5 - x_1 - x_2$$
$$x_4 = 3 - x_1$$
$$x_5 = 4 - x_2$$
$$x_6 = 3 + x_1 + x_2$$

We have reduced the original problem to the optimization problem of a function of two variables x_1 and x_2:

$$g(x_1, x_2) = f(x_1, x_2, (5 - x_1 - x_2), (3 - x_1), (4 - x_2),$$
$$(3 + x_1 + x_2)).$$

subject to the following constraints (inequalities only):

$$x_1 \geq 0, x_2 \geq 0$$
$$5 - x_1 - x_2 \geq 0$$
$$3 - x_1 \geq 0$$
$$4 - x_2 \geq 0$$
$$3 + x_1 + x_2 \geq 0$$

These inequalities can be further reduced to:

$$0 \leq x_1 \leq 3$$
$$0 \leq x_2 \leq 4$$
$$x_1 + x_2 \leq 5$$

This would complete the first step of our algorithm: elimination of equalities.

Now let us consider a single point from the search space, $\vec{x} = \langle x_1, x_2 \rangle = \langle 1.8, 2.3 \rangle$. If we try to change the value of variable x_1 without changing the value of x_2 (uniform mutation), the variable x_1 can take any value from the range: $[0, 5 - x_2] = [0, 2.7]$. Additionally, if we have two points within search space, $\vec{x} = \langle x_1, x_2 \rangle = \langle 1.8, 2.3 \rangle$ and $\vec{x}' = \langle x_1', x_2' \rangle = \langle 0.9, 3.5 \rangle$, then any linear combination $a\vec{x} + (1 - a)\vec{x}', 0 \leq a \leq 1$, would yield a point within search space, *i.e.*, all constraints must be satisfied (whole arithmetical crossover). The problem now reduces to designing such closed operators.

12.6.3 Elimination of Equalities

This processing is mechanized using matrix transformation. A detailed description is given in Reference [27].

12.6.4 Representation Issues

Let us denote the four trapezoidal corners, defining the fuzzy set v_p^i, as $\beta_1^{i,p}$, $\beta_2^{i,p}$, $\beta_3^{i,p}$, and $\beta_4^{i,p}$ (we show all the indexes only when necessary to avoid ambiguity). Then, a chromosome is a vector of such corners arranged as follows:

$$\langle corners(a_1), \ldots corners(a_n) \rangle$$

where $corners(a_i) = corners(v_1^i), \ldots corners(v_{|D_i|}^i)$, and $corners(v_p^i) = \beta_1^{i,p}$, $\beta_2^{i,p}$, $\beta_3^{i,p}$, $\beta_4^{i,p}$, each β is a gene, bounded by problem-specific domains.

12.6.5 Initialization process

A genetic algorithm requires a population of potential solutions to be initialized and then maintained during the process. For the closed operators, all initial solutions must be contained within the convex space.

12.6.6 Genetic Operators

The operators used in our system are quite different from the classical ones since to adhere to constraints some genes of a chromosome are dependent on the other genes.

The value of the i-th component of a feasible solution $\vec{s} = \langle v_1, \ldots, v_m \rangle$ is always in some (dynamic) range $[l, u]$; the bounds l and u depend on the other vector's values $v_1, \ldots, v_{i-1}, v_{i+1}, \ldots, v_m$, and the set of inequalities. We say that the i-th component (i-th gene) of the vector $\vec{s}$ is *movable* if $l < u$.

There are two important properties of convex spaces that are used here (due to the linearity of the constraints, the solution space is always a convex space S):

1. For any two points s_1 and s_2 in the solution space S, the linear combination $a \cdot s_1 + (1 - a) \cdot s_2$, where $a \in [0, 1]$, is a point in S.

2. For every point $s_0 \in S$ and any line p such that $s_0 \in p$, p intersects the boundaries of S at precisely two points, say $l_p^{s_0}$ and $u_p^{s_0}$.

Since we are only interested in lines parallel to each axis, to simplify the notation we denote by $l_{(i)}^s$ and $u_{(i)}^s$ the i-th components of the vectors l_p^s and u_p^s, respectively, where the line p is parallel to the axis i. We assume further that $l_{(i)}^s \le u_{(i)}^s$.

12.6.6.1 Mutation

Mutation is quite different from the traditional one with respect to both the actual mutation (a gene is mutated in a dynamic range) and to the selection of an applicable gene. Traditional mutation is performed on static domains for all genes. In such a case the order of possible mutations on a chromosome does not influence the outcome. This is not true anymore with the dynamic domains. To solve the problem we proceed as follows: a chromosome selected for mutation has some genes mutated in a random order.

- **uniform mutation** for this mutation we select a random gene k (from the set of movable genes of the given chromosome s determined by its current context). If $s_v^t = \langle v_1, \ldots, v_m \rangle$ is a chromosome and the k-th component is the selected gene, the result is a vector $s_v^{t+1} = \langle v_1, \ldots, v_k', \ldots, v_m \rangle$, where v_k' is a random value (uniform probability distribution) from the range $[l_{(k)}^{s_v^t}, u_{(k)}^{s_v^t}]$. The dynamic values $l_{(k)}^{s_v^t}$ and $u_{(k)}^{s_v^t}$ are easily calculated from the set of constraints (inequalities).

- **boundary mutation** is a variation of the uniform mutation with v_k' being either $l_{(k)}^{s_v^t}$ or $u_{(k)}^{s_v^t}$, each with equal probability.

- **non-uniform mutation** is one of the operators responsible for the fine tuning capabilities of the system. It is similar to uniform mutation except the probability distribution is such that v_k' is more likely to be close to v_k. Moreover, this distribution denses around v_k as the population ages.

12.6.6.2 Crossover

- **simple crossover** is defined as follows: if $s_v^t = \langle v_1, \ldots, v_m \rangle$ and $s_w^t = \langle w_1, \ldots, w_m \rangle$ are crossed after the k-th position, the resulting offspring are:

$$s_v^{t+1} = \langle v_1, \ldots, v_k, w_{k+1} \cdot a + v_{k+1} \cdot (1-a), \ldots, w_m \cdot a + v_m \cdot (1-a) \rangle \in S$$

$$s_w^{t+1} = \langle w_1, \ldots, w_k, v_{k+1} \cdot a + w_{k+1} \cdot (1-a), \ldots, v_m \cdot a + w_m \cdot (1-a) \rangle \in S$$

for some $a \in [0, 1]$. To obtain the greatest possible information exchange we need to determine the largest a such that the two offspring are indeed contained in S. This is efficiently implemented with a rough binary search.

- **single arithmetical crossover** is defined as follows: if $s_v^t = \langle v_1, \ldots, v_m \rangle$ and $s_w^t = \langle w_1, \ldots, w_m \rangle$ are to be crossed, the resulting offspring are $s_v^{t+1} = \langle v_1, \ldots, v_k', \ldots, v_m \rangle$ and $s_w^{t+1} = \langle w_1, \ldots, w_k', \ldots, w_m \rangle$, where $k \in [1, m]$, $v_k' = a \cdot w_k + (1 - a) \cdot v_k$, and $w_k' = a \cdot v_k + (1 - a) \cdot w_k$.

Here, a is a random choice from the following range:

$$
a \in \begin{cases}
[max(\frac{l^{s_w}_{(k)}-w_k}{v_k-w_k}, \frac{u^{s_v}_{(k)}-v_k}{w_k-v_k}), min(\frac{l^{s_v}_{(k)}-v_k}{w_k-v_k}, \frac{u^{s_w}_{(k)}-w_k}{v_k-w_k})] & \text{if } v_k > w_k \\
[0,0] & \text{if } v_k = w_k \\
[max(\frac{l^{s_v}_{(k)}-v_k}{w_k-v_k}, \frac{u^{s_w}_{(k)}-w_k}{v_k-w_k}), min(\frac{l^{s_w}_{(k)}-w_k}{v_k-w_k}, \frac{u^{s_v}_{(k)}-v_k}{w_k-v_k})] & \text{if } v_k < w_k
\end{cases}
$$

To increase the applicability of this operator (to ensure that a will be non–zero, which actually always nullifies the results of the operator) it is wise to select the applicable gene as a random choice from the intersection of movable genes of both chromosomes.

- **whole arithmetical crossover** is defined as a linear combination of two vectors: if s^t_v and s^t_w are to be crossed, the resulting offspring are $s^{t+1}_v = a \cdot s^t_w + (1 - a) \cdot s^t_v$ and $s^{t+1}_w = a \cdot s^t_v + (1 - a) \cdot s^t_w$. This operator uses a simpler static system parameter $a \in [0..1]$, as it always guarantees closedness.

12.7 Constraints for Optimizing Fuzzy Trees

Let us denote $\Psi = |D_1| + \cdots |D_n| + |D_C|$ to be the total number of fuzzy sets, $\eta = |A| + 1$ to be the total number of optimization attributes, and $N = 4$ to be the number of parameters needed per one trapezoidal set. Then, the total number of optimization parameters is $V = N \cdot \Psi$ (plus some additional parameters if some norms are also optimized), and the search space is $\Omega = |U|^{N \cdot \Psi}$. This number is huge for any uninformed optimization method. Moreover, most problems of interest are highly multi-modal. These two properties prohibit the use of most traditional optimization methods. That is why we selected genetic algorithms, which have been designed for such cases.

Even though the size of the search space is so huge, the solution space is actually much smaller since many subspaces of Ω are infeasible. This is where constraints begin to play an important role. However, constraints are usually not welcome with genetic algorithms. Many previous approaches to GA optimization of fuzzy sets followed the idea of implementing some of the constraints into the representation (for example, a symmetrical trapezoid can be represented by its center, lower base, and upper base, instead of the four corners). However, the remaining constraints would then be implemented as a penalty function.

In our algorithm, equality constraints are processed to explicitly reduce V by removing dependent variables. Then, inequalities are used by closed genetic operators, which produce only feasible offspring from feasible parents. This approach not only guarantees a feasible solution, but it also improves the search efficiency and is general enough to use any constraint one may need for a particular application (*e.g.*, use triangles instead of trapezoids).

In the next section we look at potential constraints to be used while optimizing trapezoidal fuzzy sets for the fuzzy decision trees.

12.7.1 Inequality Constraints

1. **Trapezoidal constraint:** $\beta_1 \leq \beta_2 \leq \beta_3 \leq \beta_4$. This constraint does not explicitly reduce the number of optimization variables V, but it reduces Ω. The total reduction is $2^{(N-1)\cdot\Psi}$.

2. **Non-containment:** $\beta_4^{p+1} > \beta_4^p$. This constraint ensures that no fuzzy set range contains the range of another fuzzy set – even though some researchers allow that. Again, this inequality does not reduce the number of optimization variables V. However, it explicitly reduces the search space Ω by $2^{(\Psi-\eta)}$.

12.7.2 Equality Constraints

1. **Overlap:** $y \cdot \beta_3^p + (1-y) \cdot \beta_4^p = (1-y) \cdot \beta_1^{p+1} + y \cdot \beta_2^{p+1}$. This constraint applies to neighboring fuzzy sets and deals with overlaps. The overlap is defined here as the membership value of an input which evaluates the same in two adjacent sets. In many practical applications, this overlap is set to 1/2. When this is the case (or in general when $y = const$), the constraint becomes an equality. This strong case ($y = const$) reduces V by $\Psi - \eta$ and Ω by $|U|^{\Psi-\eta}$. In the weak case ($y > 0$), this constraint reduces to the inequality $\beta_4^p > \beta_1^{p+1}$. This, in turn, reduces the search space by only $2^{\Psi-\eta}$.

2. **Completeness:** $(\beta_2^1 = 0) \wedge (\beta_3^{|D_i|} = 1)$. This constraint states that the leftmost and the rightmost fuzzy sets must completely cover the left and right portions of U (for completeness). This translates to reduction of two variables per attribute and decision. That is, V is reduced by 2η, and Ω is reduced by $|U|^{2\eta}$.

3. **Set symmetry:** $\beta_2 - \beta_1 = \beta_4 - \beta_3$. This constraint states that the trapezoids/triangles are symmetrical. It may not always be desired. However, when used it can reduce the number of optimization variables V by Ψ, reducing Ω by $|U|^\Psi$.

4. **Domain symmetry:** $\beta_m^i = \beta_m^{|D_i|-1}$, for $m = 1..4$. This constraint is often used when the fuzzy variable contains labels for fuzzy sets on both sides of a zero value and it is assumed that the properties are symmetrical, as in $Signal = \{Negative, Zero, Positive\}$. It reduces each domain D_i by $\lfloor |D_i| \rfloor$, or Ψ by up to $\Psi/2$. This reduces the search space by up to $|U|^{2\Psi}$.

5. **Triangular**: $\beta_2 = \beta_3$. This constraint changes all fuzzy sets to triangles. This might be beneficial due to the search space reduction if no representative power is lost. Most experiments indicate that both triangles and trapezoids have similar representative power. This constraint reduces the number of optimization variables by Ψ, which translates to Ω reduction of $|U|^{\Psi}$.

6. **Definition-sharing**: $\beta_m^i = \beta_m^j$, for some attributes a_i and a_j, and $m = 1..4$. This constraint states that some attributes referring to the same U may share fuzzy sets for their linguistic terms. This not only increases comprehensibility. It also dramatically reduces Ψ, which subsequently reduces Ω. For example, this constraint reduces the initial search space, before any other constraints are used, to $\Omega = |U|^{N \cdot \Psi / \eta}$.

It is important to note that not all of the constraints are independent. There are some obvious redundancies. For example, the definition-sharing constraint reduces the number of optimization parameters, thus affecting the impact of any other constraint. Also, we obviously have the capability to specify the constraints selectively for specific sets or attributes.

In addition, this optimization algorithm for fuzzy decision trees introduces its own constraints of the form $a_i = const$. These constraints can be easily incorporated with our algorithm by automatically removing the appropriate genes from all chromosomes. However, to avoid modifications of the existing algorithm, the same can be accomplished by setting singleton domains.

12.8 An Illustration

As an illustration, we extended an experiment reported in Reference [41]. However, our experiment is designed to illustrate optimization capabilities of the fuzzy tree, and the savings generated by utilizing various constraints. Other experiments will be presented in the future. This experiment involves learning the *Mexican Sombrero* function, illustrated in Figure 12.4a, from a set of data points. In the original experiment, 13 neural networks were used to incorporate predefined fuzzy rules, spanned by predefined fuzzy sets. There were 13 linguistic labels, and thus 13 fuzzy sets, per attribute, and there were 7 fuzzy sets for function values (decisions). Figure 12.4b and 12.4c illustrate the mapping as acquired by our fuzzy tree using two different inferences—the second inference makes the tree similar to a symbolic decision tree.

We extended this example to ten dimensions, appropriately increasing the number of training data points. Therefore, $\Psi = 137$. Assuming 0.01 precision, and assuming that the fuzzy sets were trapezoids, $V = 549$, $\Omega = 101^{549} \approx 10^{1250}$.

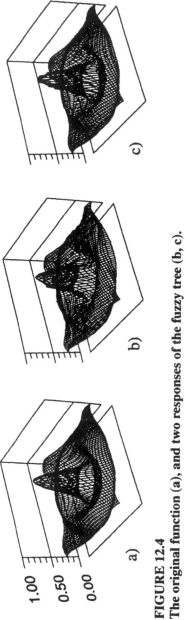

FIGURE 12.4
The original function (a), and two responses of the fuzzy tree (b, c).

We started with exactly the same fuzzy sets as those used in Reference [41], and we ran our modified tree-building procedure of Sections 12.5 and 12.6 without any constraints, but instead penalizing infeasible solutions. We iterated adjusting the penalty until feasible solutions were being consistently generated. After the penalty was fixed, we repeated the run, measuring t_0 as the time needed for the static optimization to saturate with some I_0 as the minimal information measure. Then, we continued the tree-building procedure with dynamic optimization. Because the search space decreases in dynamic optimization, we were allowing optimization time $t_0/ \mid A^N \mid$ at a node N. At the end, we measured the quality of the generated fuzzy tree by measuring sum of squared errors between the represented and the actual function on a dense grid. Let us denote this error E_0.

We subsequently restarted with the same fuzzy sets, this time allowing the inequality, and the first two equality, constraints. This experiment was conducted in two independent settings. Note that the penalty was not needed any more.

1. We allowed exactly the same optimization times as previously. This resulted in 8% improvement in I_0 and 11% improvement in E_0.

2. We measured a new time t_1 needed for the same final information contents I_0 in the static optimization, and speeded up the dynamic optimization by using now $t_1/ \mid A^N \mid$. The error measure improved by 5% over E_0, and the total procedure time was reduced by 26%.

Next, we allowed the next three additional equality constraints. Again, we repeated two new independent tree-building procedures in exactly the same settings as previously.

1. I_0 was reduced by 20% and E_0 by 29%.

2. E_0 improved by 8% in total time which shortened by 72%.

Finally, we also used the definition-sharing constraint. This reduces the initial search space to about 10^{162}, a sizable improvement. This constraint was applicable in this experiment since the "Sombrero" function is symmetrical. Again, we repeated the two independent settings.

1. I_0 was reduced by 22% and E_0 by 69%.

2. E_0 was reduced by 54%, in time shortened by 90%.

12.9   Conclusions

We presented an optimization method for fuzzy trees. This method optimizes the fuzzy component of a fuzzy tree, and it is incorporated with the fuzzy-tree building routine reported previously. We also presented a particular implementation of the method, based on GAs for processing linear constraints. We identified all constraints that can be used to reduce the search space, along with expected reductions. Finally, we presented a simple experiment illustrating the improvements gained in both time needed for the optimization as well as in the net result.

In the future, we plan to explore optimization of the tree structure. We also hope to combine that future optimization with the current tree-building routine, so that both kinds of optimization could be performed simultaneously.

References

[1] Baires, E.R., Porter, B.W., and Wier, C.C., PROTOS: an exemplar-based learning apprentice, in *Machine Learning III*, Kondratoff, Y. and Michalski, R., Eds., Morgan Kaufmann, San Mateo, CA, 112.

[2] Breiman, L., Friedman, J.H., Olsen, R.A., and Stone, C.J., *Classification and Regression Trees*, Wadsworth, 1984.

[3] Boverie, S. et al., Fuzzy logic control compared with other automatic control approaches, in *Proc. 30th Conf. on Decision and Control*, 1212.

[4] Buckley, J.J. and Ying, H., Expert fuzzy controller, *Fuzzy Sets Syst.*, 44, 373, 1991.

[5] Clark, P. and Niblett, T., Induction in noisy domains, in *Progress in Machine Learning*, Bratko, I. and Lavrac, N., Eds., Sigma Press, 1987.

[6] Davis, L., Ed., *Handbook of Genetic Algorithms*, Van Nostrand Reinhold, New York, 1991.

[7] Dietterich, T.G., Hild, H., and Bakiri, G., A comparative study of ID3 and backpropagation for English text-to-speech mapping, in *Proc. Int. Conf. on Machine Learning*, 1990.

[8] Enbutsu, I., Baba, K., and Hara, N., Fuzzy rule extraction from a multilayered neural network, in *Int. Joint Conf. on Neural Networks*, 1991, 461.

[9] Fisher, D.H., Knowledge acquisition via incremental conceptual clustering, in *Machine Learning*, Vol. 2, Kluwer Academic Publishers, Boston, 1987.

[10] Goldberg, D.E., *Genetic Algorithms in Search, Optimization, and Machine Learning*, Addison-Wesley, Reading, MA, 1989.

[11] Holland, J., *Adaptation in Natural and Artificial Systems*, University of Michigan Press, Ann Arbor, 1975.

[12] Janikow, C.Z., Learning fuzzy controllers by genetic algorithms, in *Proc. 1994 ACM Symp. on Applied Computing*, ACM Press, New York, 1994.

[13] Janikow, C.Z., Fuzzy processing in decision trees, in *Proc. 6th Int. Symp. on Artificial Intelligence*, Noriega Megabyte, Mexico, 1993, 360.

[14] Janikow, C.Z., Fuzzy decision trees: FIDMV, in *First Proc. Joint Conf. on Information Sciences—Fuzzy Theory and Technology*, Paul P. Wang, Ed., Pinehurst, NC, 1994, 232–235.

[15] Janikow, C.Z., Fuzzy decision trees: issues and methods, *IEEE Trans. Syst. Man. Cybern.*, to appear.

[16] Janikow, C.Z., A knowledge-intensive genetic algorithm for supervised inductive learning in attribute-based spaces, *Machine Learning*, 13(2/3), 189, 1993.

[17] Janikow, C.Z., A genetic algorithm method for optimizing fuzzy decision trees, *Inf. Sci.*, to appear.

[18] Janikow, C.Z., *User's Manual for FIDMV*, available by anonymous ftp from radom.umsl.edu, in pub/FIDMV.

[19] Kandel, A. and Langholz, G., Eds., *Fuzzy Control Systems*, CRC Press, Boca Raton, FL, 1993.

[20] Katayama, R., Kajitani, Y., and Nishida, Y., A self generating and tuning method for fuzzy modeling using interior penalty method and its application to knowledge acquisition of fuzzy controller, in *Fuzzy Control Systems*, CRC Press, Boca Raton, FL, 1993.

[21] Kobayashi, K., Ogata, H., and Murai, R., Method of inducing fuzzy rules and membership functions, in *Proc. IFAC Symp. on Artificial Intelligence in Real-Time Control*, 1992, 283.

[22] Lebowitz, M., Categorizing numeric information for generalization, *Cognitive Sci.*, 9, 285, 1985.

[23] Liepins, G.E. and Potter, W.D., A genetic algorithm approach to multiple-fault diagnosis, in *Handbook of Genetic Algorithms*, Van Nostrand Reinhold, New York, 1991.

[24] McNeill, D. and Freiberger, P., *Fuzzy Logic*, Simon & Schuster, New York, 1993.

[25] Michalewicz, Z. and Janikow, C.Z., Genetic algorithms for numerical optimization, *Stat. Comput.*, 1, 75, 1991.

[26] Michalewicz, Z. and Janikow, C.Z., Handling constraints in genetic algorithms, in *Proc. 4th Int. Conf. on Genetic Algorithms*, Morgan Kaufmann, San Mateo, CA, 1991, 151.

[27] Michalewicz, Z. and Janikow, C.Z., GENOCOP: a genetic algorithm for numerical optimization problems with constraints, *Commun. ACM*, to appear.

[28] Michalski, R.S., A theory and methodology of inductive learning, in *Machine Learning I*, Michalski, R.S., Carbonell, J.G., and Mitchell, T.M., Eds., Morgan Kaufmann, San Mateo, CA, 1986, 83.

[29] Michalski, R.S., Learning flexible concepts, in *Machine Learning III*, Michalski, R. and Kondratoff, Y., Eds., Morgan Kaufmann, San Mateo, CA, 1991.

[30] Michalski, R.S., Understanding the nature of learning, in *Machine Learning: An Artificial Intelligence Approach*, Vol. 2, Michalski, R., Carbonell, J., and Mitchell, T., Eds., Morgan Kaufmann, San Mateo, CA, 1986.

[31] Mingers, J., An empirical comparison of pruning methods for decision tree induction, in *Machine Learning*, Vol. 4, Morgan Kaufmann, San Mateo, CA, 1989, 227.

[32] Mingers, J., An empirical comparison of selection measures for decision-tree induction, in *Machine Learning*, Vol. 3, Morgan Kaufmann, San Mateo, CA, 1989, 319.

[33] Mizumoto, M., Fuzzy controls under various fuzzy reasoning methods, in *Inf. Sci.*, 45, 129, 1988.

[34] Quinlan, J.R., The effect of noise on concept learning, in *Machine Learning II*, Michalski, R., Carbonell, J., and Mitchell, T., Eds., Morgan Kaufmann, San Mateo, CA, 1986.

[35] Quinlan, J.R., Induction on decision trees, in *Machine Learning*, Vol. 1, Morgan Kaufmann, San Mateo, CA, 1986, 81.

[36] Quinlan, J.R., Decision trees as probabilistic classifiers, in *Proc. 4th Int. Workshop on Machine Learning*, Morgan Kaufmann, Los Altos, CA, 1987.

[37] Quinlan, J.R., Unknown attribute-values in induction, in *Proc. 6th Int. Workshop on Machine Learning*, Morgan Kaufmann, San Mateo, CA, 1989.

[38] Quinlan, J.R., *C4.5: Programs for Machine Learning*, Morgan Kaufmann, San Mateo, CA, 1993.

[39] Richardson, J.T., Palmer, M.R., Liepins, G., and Hilliard, M., Some guidelines for genetic algorithms with penalty functions, in *Proc. 3rd. Int. Conf. on Genetic Algorithms*, Morgan Kaufmann, San Mateo, CA, 1989.

[40] Sestino, S. and Dillon, T., Using single-layered neural networks for the extraction of conjunctive rules and hierarchical classifications, *J. Appl. Intelligence 1*, 157, 1991.

[41] Suh, I.H. and Kim, T.W., Fuzzy membership function based on neural networks with applications to the visual serving of robot manipulators, *IEEE Trans. Fuzzy Syst.*, 2(3), 203,1994.

[42] Wang, L. and Mendel, J., Generating fuzzy rules by learning from examples, *IEEE Trans. Syst. Man Cybern.*, 22(6), 1414, 1992.

13

Genetic Design of Fuzzy Controllers

Mark G. Cooper and Jacques J. Vidal

Abstract   This chapter considers the application of genetic algorithms as a general methodology for automatically generating fuzzy process controllers. In contrast to prior genetic-fuzzy systems which require that every input-output combination be enumerated, we propose a novel encoding scheme which maintains only those rules necessary to control the target system. The key to our method, demonstrated on the classic cart-pole problem, is to represent each fuzzy system as an unordered list of an arbitrary number of rules. Both the composition and size of the rule base evolve from an initial random state. By using a rule-based representation, successful systems evolve quickly, increasing the likelihood of success when applied to more complex problems.

13.1   Introduction

The ability to translate information supplied in linguistic terms into a computer-usable form has made fuzzy logic systems popular for implementing complex process control. However, most only represent the "best guesses" of human experts and can therefore be improved using feedback to *tune* the initial fuzzy rule bases. Automatic rule-base generation has been attempted using gradient descent techniques, feedforward neural networks, radial basis function networks, fuzzy clustering, and genetic algorithms. These systems learn to control the targeted process through training—examples are presented and the system is modified based on performance evaluation. Unfortunately, the number of rules must generally be specified *a priori,* and learning capacity is often small. We believe, however, that these limitations are not inherent to the problem, but rather lie in the details of the learning procedure employed.

To overcome these deficiencies, our approach uses a genetic algorithm to determine the number of rules in the fuzzy rule base as well as their composition. Additionally, a novel encoding scheme for the fuzzy rules results in a more compact rule base than used previously, allowing more complexity to evolve.

We begin by introducing the fuzzy inference model, followed by a brief description of genetic algorithms (for a complete treatment of genetic algorithms, see Goldberg [2]), and of the physical system used for demonstration—the cart-pole problem. Following a review of two earlier genetic-fuzzy proposals, we describe our alternative approach, present experiment results, and discuss the properties and limitations of our technique.

13.1.1   The Fuzzy Associative Memory Model

Fuzzy rules typically appear in the form "If x_1 is v_1 and x_2 is v_2 and ..., then a_1 is w_1 and a_2 is w_2 and ...," where x_n are input variables, a_n are output variables, and v_n and w_n are descriptors like "small positive", "large negative", "medium", etc. Associated with each descriptor is a *membership function* specifying the degree to which an input satisfies the descriptor. A *Fuzzy Associative Memory* (or FAM) [4] applies these rules to a set of inputs, combines the consequents of each rule, and produces a value for each output variable. To illustrate the operation of a fuzzy associative memory, we present two rules from a simple hypothetical fuzzy system for controlling an automobile cruise control. The controller inputs are the difference between the current speed and the desired speed (s) and the acceleration (a) of the car, and the output is the recommended throttle adjustment (t).

1.   If the difference between the current speed and the desired speed is a *small positive* value, and the acceleration is *near zero,* then adjust the throttle by a *small negative* value.

2.   If the difference between the current speed and the desired speed is *near zero,* and the acceleration is a *small negative* value, then adjust the throttle by a *small positive* value.

These rules are displayed graphically in Figure 13.1. The top row corresponds to the first rule, and the bottom corresponds to the second. The first two columns represent the antecedent clauses, while the third represents the consequent clause.

Instantiations of the input variables are presented to the rule base in parallel. The membership function for each antecedent clause is applied to its corresponding input to produce a fit value for the clause. When the rule antecedent is made up of a conjunction of clauses, the minimum fit value among the clauses is selected as the fit for the entire rule, i.e., as the degree of applicability of the rule to the overall inference. In the example, the input values $s = +3$ mph and $a = -1$ mph/s are applied to the rule base producing antecedent fit values of 1.0 and 0.8 for the

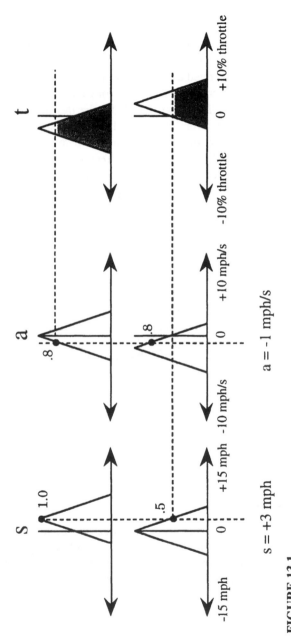

FIGURE 13.1
Fuzzy reasoning.

first rule, and 0.5 and 0.8 for the second. The minimum of each pair specifies the overall fit for each rule.

Next, the consequent of each rule is calculated as the *region* under its membership function below the antecedent fit value. The output of the first rule is the region of the consequent's membership function below the antecedent fit value 0.8. Similarly, the output of the second rule is the region under 0.5.

The collection of regions generated as the output of the fuzzy rules is the qualified output of the FAM. To generate a single value for each consequent variable, these regions must be combined. This process, called *defuzzification* (Figure 13.2), may be accomplished by any of several methods. Each method yields a slightly different result given the same set of output regions; however, these differences are inconsequential when viewed in the context of effecting ongoing process control, and, in a system based on learning, are implicitly integrated when the fuzzy controller is trained. In other words, the training procedure will discover an appropriate set of rules regardless of the defuzzification method selected. For our purposes, we compute the average of the centers of the output regions weighted by their areas.

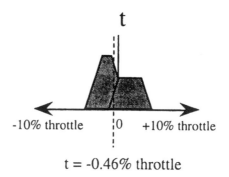

$t = -0.46\%$ throttle

FIGURE 13.2
Defuzzification.

13.1.2   Genetic Algorithms

Genetic algorithms are probabilistic search techniques that emulate the mechanics of evolution. They are capable of globally exploring a solution space, pursuing potentially fruitful paths while also examining random points to reduce the likelihood of settling for a local optimum. The system being evolved is encoded into a long bit string called a *chromosome*. Sites on the chromosome correspond to specific characteristics of the encoded system and are called *genes*. Initially, a random set, or *population*, of these strings is generated. Each string is then evaluated according to a given performance criterion and assigned a fitness score. The strings with the best scores are used in the reproduction phase to produce the next generation.

The reproduction of a pair of strings proceeds by copying bits from one string until a randomly triggered *crossover* point, after which bits are copied from the other string. As each bit is copied, there is also the probability that a *mutation* will occur. Mutations include inversion of the copied bit and the addition or deletion of an entire rule (Figure 13.3). These latter two mutations permit the size of a system's rule base to evolve. The cycle of evaluation and reproduction continues for a predetermined number of generations, or until an acceptable performance level is achieved.

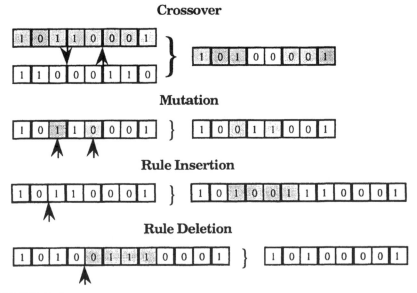

FIGURE 13.3
Genetic operators.

13.1.3 Pole Balancing

We will apply the evolved fuzzy systems to a classic control problem variously referred to as the "cart-pole", "inverted pendulum", or "pole balancing" problem. This problem is an example of an inherently unstable dynamic system, and is an established literature benchmark for evaluating control schemes. The objective is to control translational forces that position a cart at the center of a finite width track while simultaneously balancing a pole hinged on the cart's top (Figure 13.4). The physical plant can be simulated from a set of nonlinear differential equations describing the cart and pole dynamics [6].

$$\ddot{x} = \frac{F - \mu_c sign(\dot{x}) + \tilde{F}}{M + \tilde{m}} \tag{13.1}$$

$$\ddot{\theta} = -\frac{3}{4l}\left(\ddot{x}\cos\theta + g\sin\theta + \frac{\mu_p\dot{\theta}}{ml}\right) \qquad (13.2)$$

where

$$\tilde{F} = ml\dot{\theta}^2\sin\theta + \frac{3}{4}m\cos\theta\left(\frac{\mu_p\dot{\theta}}{ml} + g\sin\theta\right) \qquad (13.1a)$$

$$\tilde{m} = m\left(1 - \frac{3}{4}\cos^2\theta\right) \qquad (13.1b)$$

Table 13.1 Symbols Used in Equations 13.1 and 13.2

Symbol	Description	Value
x	Cart position	$[-1.0, 1.0]$ m
θ	Pole angle from vertical	$[-0.26, 0.26]$ rad
F	Force applied to cart	$[-10, 10]$ N
g	Force of gravity	9.8 m/s^2
l	Half-length of pole	0.5 m
M	Mass of the cart	1.0 kg
m	Mass of the pole	0.1 kg
μ_c	Friction of the cart on track	0.0005 N
μ_p	Friction of pole's hinge	0.000002 kg m^2

The cart and pole are initially placed at rest at a predetermined position. A simulation succeeds when 60 s of simulated time elapses without the cart reaching the end of the track or the pole falling over. For our purposes, we consider a pole angle of 0.26 rad (about 14°), the point at which the pole "falls over." Wieland [6] investigated the effect of varying the angle at which a failure occurs on the process of evolving a neural network that controls the cart-pole system and concluded that increasing this parameter does not necessarily generate superior solutions and makes training slower.

13.2   Generating Fuzzy Systems Using Genetic Algorithms

A central issue in fuzzy control is the selection and encoding of the fuzzy rules. Sufficient system information must be encoded and the representation must be

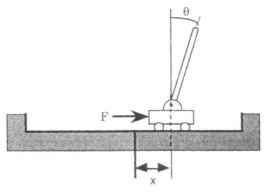

FIGURE 13.4
The cart-pole system.

amenable to evaluation and, in the genetic framework, reproduction. The representation must allow flexibility both in the number and content of the discovered rules. Finally, a proper scoring function is essential. We will describe two earlier applications of genetic algorithms to the automatic generation of a fuzzy rule base and compare these to our approach.

One of the earliest applications of genetic algorithms to the design of fuzzy systems was developed by Karr [3], again to solve the cart-pole problem. The production of the fuzzy controller begins with the definition of the fuzzy sets used to describe each input variable. Each of the four input variables are characterized by three fuzzy sets—NEGATIVE, ZERO, and POSITIVE—yielding 81 possible combinations. The fuzzy system designer then assigns one of seven choices for the output to each input combination. The resulting fuzzy system represents the expert's "best guess". The membership function extrema are then encoded into a bit string, and a genetic algorithm is applied to shift the membership functions so as to find locations which improve performance. The evolved system consistently outperforms the original, being capable of recovering from initial positions that fail under the original rule base.

Lee and Takagi [5] improved on Karr's method by permitting both the size and the composition of the rule base to emerge as the result of the genetic search, instead of being specified by the system designer. However, the maximum number of fuzzy sets permitted for each variable is still fixed. Each membership function specification consists of center, right base, and left base values, each requiring one byte of storage. Each output variable has a three-byte description of the action to be taken in response to every input variable combination. There are $\prod m_i$ descriptions associated with each output variable, where m_i is the maximum number of membership functions permitted for input variable i. Therefore, the byte length of the entire string is $3 \left(\sum m_i + \nu \prod m_i \right)$, where ν is the number of output variables. In their example, each input variable is assigned a maximum of ten membership functions, and each system description requires 360 bytes. Hence, 30,120 bytes are required to encode the four-input version of the problem. As the number of variables increases the length of the string encoding the system size increases exponentially, with a corresponding exponential increase in the complexity of the search space. Such a system is unlikely to scale well for more complex problems.

13.3   A Rule-Based Approach

13.3.1   System Representation

It is a long-recognized limitation of genetic algorithms that, as a global search technique, the encoded bit string lengths should be kept as small as possible as they evolve, since the size of the search space increases exponentially with the size of the strings. Therefore, it is highly desirable to maintain only the rules necessary to accomplish the task. This is the goal pursued in our approach. The two evolutionary requirements listed in Lee and Takagi [5] are also maintained; namely, that the number of rules needed to accomplish the task must be evolved, and that the membership functions comprising those rules must evolve from a random initial state.

Our version of the cart-pole problem has four input variables (x-position, θ-position, dx/dt, $d\theta/dt$) and one output variable (the force applied to the cart). The membership function for each variable is a triangle characterized by the location of its center and the half-length of its base. A single rule, therefore, consists of the concatenation of ten one-byte unsigned characters (assuming values from 0 to 255) specifying the centers and half-lengths of the membership functions (Figure 13.5). The rule descriptions for a single fuzzy system are then concatenated into a single bit string where the number of rules is not restricted.

One limitation of earlier attempts at fuzzy system learning was the assumption that each rule is dependent on combinations of all of the input variables. In many cases this is undesirable since the appropriate action is often strongly conditioned by only a small subset of the variables. Without the ability to ignore irrelevant variables, rules evolve which account for every combination of spurious variables with each required value for the relevant ones. We address this problem by ignoring a variable when its corresponding half-length falls outside of a particular range ([40, 215] out of the overall range of [0, 255] in our experiments).

13.3.2   System Evaluation

Each evolved fuzzy system is evaluated according to the length of time that it keeps the cart-pole system from failing. Twenty different initial starting positions are considered, arranged symmetrically around $\{x = 0, \theta = 0\}$ (i.e., if the position

Fuzzy System Population

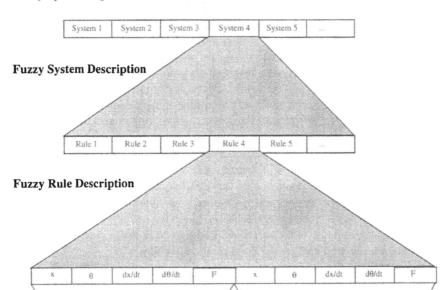

FIGURE 13.5
Organization of the fuzzy system population.

$\{x = 0.5, \theta = 0.07\}$ is in the test suite, so too are $\{0.5, -0.07\}$, $\{-0.5, -0.07\}$, and $\{-0.5, 0.07\}$). The selection of a test suite representative of all possible combinations of input values is critical for evolving a robust system.

Each fuzzy controller's objective is to keep the pole balanced and the cart positioned near the center of the track. The score for a particular trial accumulates as long as the system does not fail. Once a failure occurs the next test immediately begins. Performance scores, however, need only be based on the deviation of cart's position from the center. The pole's angle is not required explicitly in the scoring equation since it is already an implicit constraint on the score—once the pole angle exceeds a critical value the simulation is terminated. In fact, when both the x-position and angle are included in the scoring equation, the location of the cart is almost entirely ignored, whereas including only the x-position in the scoring equation, both variables assume comparable importance.

$$score = \sum_{\text{20 test cases}} \sum_{\text{60 s}} \sum_{100 \frac{\text{updates}}{\text{second}}} \left(1.0 - \frac{x}{x_{\max}}\right) \qquad (13.3)$$

13.3.3   Rule Alignment and Reproduction

To be meaningful, the genetic paradigm requires that the rules in the two strings be aligned so that similar rules are combined with each other. Consider, for example, two fuzzy systems having the same rule set, but arranged differently in each. Simply combining the strings in the order they appear does not preserve much information about either system and produces nearly random results, rather than a child system that performs in a manner similar to its parents. Therefore, before reproduction, the two strings must be aligned so that the centers of the input variables match as closely as possible. In our algorithm, the most closely matching rules (defined as the sum of the differences between the input variable centers) are combined first, followed by the next most closely matching rules from those that remain, and so on (Figure 13.6). Any rules from a longer string that are not matched are added at the end. This matching scheme is not optimal, but requires considerably less time and performs nearly as well as the optimal method which involves generating the difference matrix between each pair of rules, then solving the "distribution problem" to find the minimal matching.

13.4   Simulation Results

To demonstrate our training procedure, several fuzzy cart-pole controllers were evolved. The population size in each trial was 200, with each system randomly assigned between 1 and 30 rules initialized with random characters. The fitness of each controller was then established according to its performance on the cart-pole system for 60 simulated seconds, with the state of the system updated using Euler's method 100 times every second. Pairs of strings from the 50 top scoring controllers were randomly selected and combined to produce the next generation.

The probabilities with which the genetic operators were applied are permitted to vary uniformly within the following ranges: crossover [0, 0.0094] per bit copied, mutation [0, 0.0094] per bit copied, add-rule mutation [0, 0.0125] per rule, delete-rule mutation [0, 0.0125] per rule. This method for setting the genetic parameters is unique to our approach and has been found to produce better results than single-value parameters [1].

Learning typically occurred quickly. By about the 50th generation the top-scoring controller was usually capable of maintaining the cart-pole system for the entire 60 s (as opposed to the 5000 generations required in Lee and Takagi [5]). The progress of training for several runs appears in Figure 13.7.

To illustrate the results of a typical run in detail, an evolved rule base appears in Figure 13.8. Each row corresponds to a single rule, and each column corresponds to a system variable—$x, \theta, dx/dt, d\theta/dt$, and F, respectively. This rule base was then applied to a random initial position not used during training

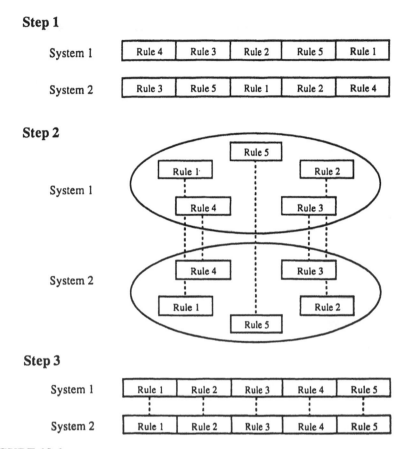

FIGURE 13.6
Rule alignment.

$(x = 0.54, \theta = -0.077)$. Traces of the x-position, θ-position, and force applied appear in Figure 13.9. The controller does not actually bring the system to rest, but rather establishes a stable oscillation. In fact, the fuzzy controller maintains the cart-pole system upright through 24 h of simulated time. It is impossible to bring the cart-pole system completely to rest (the oscillations continue to diminish). In our example, the closest that the system comes to being centered at rest after 24 h was $x = 0.000067, \theta = 0.00085, dx/dt = 0.000011$, and $d\theta/dt = 0.000027$.

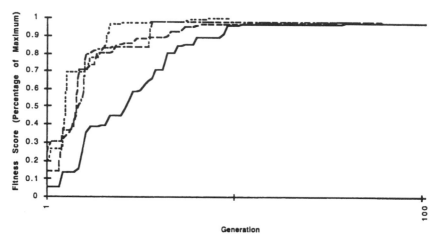

FIGURE 13.7
Progress of training for several evolutionary trials.

13.5   Discussion

The most significant contribution of our approach is the ability to evolve a fuzzy rule base consisting solely of relevant rules, in contrast to previous approaches which require that every possible combination of a fixed set of inputs be enumerated, or to require that the relevancy of a large suite of rules be determined.

The fuzzy controllers evolved by our method, like most systems generated using genetic algorithms, are very sensitive to the selection pressures imposed by the system designer. These include the training set and the performance criteria. With regard to the composition of the training set, the range and distribution of the test cases must be representative of the expected system input. If the test cases are unevenly distributed, e.g., most of the initial positions have positive x-positions while very few have negative x-positions, then the evolved systems will fail to learn the correct control action for the rarer cases. In fact, continued training after learning to correctly handle the common cases still fails to improve performance. This is because in early generations the systems performing correctly on the common cases are more likely to reproduce because they perform well on a higher percentage of the training set. In later generations, the evolved systems cannot change sufficiently to successfully handle the rarer cases without degrading performance to the point where they are not selected to reproduce. Hence, the rarer cases are never adequately learned.

Additionally, careful consideration must be made in choosing a performance criterion. In trials where the divergence of both x and θ were considered in the scoring equation, rule bases were generated which successfully balanced the pole,

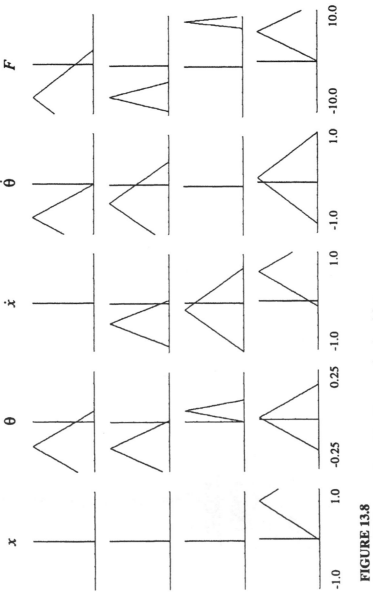

FIGURE 13.8
Fuzzy rule base produced by the genetic algorithm.

Time Steps (0.01 of a second)

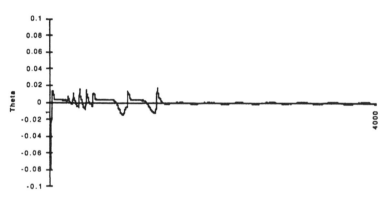

Time Steps (0.01 of a second)

Time Steps (0.01 of a second)

FIGURE 13.9
Traces of selected variables in the trial run.

but which ignored the x-position of the cart. Since success in balancing the pole is the primary prerequisite for the continuance of each test case, the x-position became an unimportant evolutionary factor and was often ignored. Therefore, we used the x-position as an explicit training criterion, while the θ-position remained an implicit training criterion.

Despite the fact that the rule bases generated by this method include only those rules required to solve the problem, there are instances where rules are duplicated, or where one rule is subsumed by another. One direction for continued research is the automatic identification of instances where rules can be combined or eliminated. This will allow the rule bases to be further consolidated. Due to the compactness of the representations, we expect that this method will be capable of addressing more complex problems than previous approaches. Future research will more fully characterize this method by observing the effects of changing system parameters (e.g., pole length) during evolution and by examining in detail the problem-solving methodologies discovered by the genetic search. More difficult problems such as the multiple pole and the jointed pole systems will be attempted in order to expand the limits of genetic fuzzy system generation.

In short, the fuzzy systems produced by genetic search are extremely faithful to both the training set and the performance criteria used during evolution. In many cases, the resulting systems perform in a manner consistent with these parameters, yet still differ from the vision of the system designer. Therefore, although the objective of automatic fuzzy system generation is to eliminate the need for a human expert by learning solely through observation and feedback, an expert is still required to shape the progress of learning and to interpret and evaluate the resulting systems.

References

[1] Cooper, M.G., Genetic Design of Rule-Based Fuzzy Controllers, Ph.D. dissertation, Department of Computer Science, University of California, Los Angeles, 1994.

[2] Goldberg, D.E., *Genetic Algorithms in Search, Optimization, and Machine Learning,* Addison-Wesley, Reading, MA, 1989.

[3] Karr, C.L., Design of a cart-pole balancing fuzzy logic controller using a genetic algorithm, *Proc. SPIE Conf. on the Applications of Artificial Intelligence,* International Society for Optical Engineering, Bellingham, WA, 1991, 26.

[4] Kosko, B., *Neural Networks and Fuzzy Systems: A Dynamical Systems Approach to Machine Intelligence,* Prentice-Hall, Englewood Cliffs, NJ, 1992.

[5] Lee, M.A. and Takagi, H., Integrating design stages of fuzzy systems using genetic algorithms, in *Proc. 2nd IEEE Int. Conf. on Fuzzy Systems,* Institute of Electrical and Electronics Engineers, New York, 1993, 612.

[6] Wieland, A.P., Evolving controls for unstable systems, in *Connectionist Models: Proc. 1990 Summer School,* Touretzky, S. et al., Eds., Morgan Kaufmann, San Mateo, CA, 1991, 91.

Index

Milton Keynes UK
Ingram Content Group UK Ltd.
UKHW031143141024
449569UK00024B/1108